AF380103

Georg Will

Powder Diffraction

The Rietveld Method and the Two Stage Method to
Determine and Refine Crystal Structures from
Powder Diffraction Data

Georg Will

Powder Diffraction

The Rietveld Method and the Two Stage Method
to Determine and Refine Crystal Structures from
Powder Diffraction Data

With 94 Figures and 43 Tables

 Springer

Author

Prof. Dr. Georg Will

Mineralogisch-Petrologisches Institut und Museum der Universität Bonn
Poppelsdorfer Schloß
53115 Bonn, Germany
or:
Schlehenweg 17
53913 Swisttal-Buschhoven, Germany
Phone: +49 (0) 2226-2939

Library of Congress Control Number: 2005931757

ISBN-10   3-540-27986-7 **Springer Berlin Heidelberg New York**
ISBN-13   978-3-540-27986-0 **Springer Berlin Heidelberg New York**

Springer is a part of Springer Science+Business Media
springeronline.com
© Springer-Verlag Berlin Heidelberg 2006
Printed in Germany

Cover design: E. Kirchner, Heidelberg
Typesetting: Büro Stasch (*stasch@stasch.com*) · Uwe Zimmermann, Bayreuth
Production: Agata Oelschläger
Printing and Binding: Stürtz, Würzburg

Printed on acid-free paper   30/2132/AO – 5 4 3 2 1 0

Preface

After the pioneering experiments by M. v. Laue, Friedrich and Knipping in 1912, performed with single crystals (for a comprehensive historical remembrance see Ewald 1962), experiments with poly-crystalline specimens, with "powders", followed almost immediately. Only four years later in 1916 the first such experiments were reported by Debye and Scherrer (Debye and Scherrer 1916a,b): "*Interferenzen an regellos orientierten Teilchen im Röntgenlicht*". Naturally, the first structures solved were rather simple compounds, LiF, silicon, or iron, but "the method has opened up a new field of the greatest interest and importance". Nevertheless "powder diffraction" drew little attention, and the application for structure determination was limited. For many years X-ray powder diffraction was used and was best known for phase analysis of crystalline mixtures according to Hanawalt, based on the so-called Hanawalt files (Hanawalt et al. 1938). This dates back to the 1930s.

The situation changed in the late 1940s. Scientist working in neutron diffraction realized the inherent potentials of powder diffraction (Shull and Smart 1949). They used powder diffraction techniques for one reason because single crystals of the required dimensions were hardy available. Secondly because only neutron diffraction offered the possibility to determine magnetic structures. With steadily improving instrumentation powder diffraction became more and more attractive and it was only natural to use powder diffraction patterns also for the purpose to derive crystal structures from these data. This dates to the beginning of the 1960s.

The inherent potential of powder diffraction for crystallographic problems was realized at first again for the analysis of neutron diffraction data. Scientists began to develop methods for using their powder diffraction data, at the beginning only for the refinement of crystal structures. The first publications appeared by Rietveld (1969), working in Petten, The Netherlands, with "*A Profile Refinement Method for Nuclear and Magnetic Structures*", and by Will (Will et al. 1965), working in Brookhaven, USA with "*The crystal structure of MnSO$_4$*", the first application of the Two Stage method. With the development of steadily growing computer power the analysis was extended to profile fitting and pattern decomposition, which allowed one to extract individual intensities from overlapping diffraction peaks. This moved powder diffraction into the neighborhood and even in competition to single crystal structure analysis.

Recently developed analysis methods have revolutionized powder diffraction. Today it has become such a powerful method that it is being used even in place of single crystal methods in many cases, even when single crystals are available. This is because of the speed and simplicity of data collection by powder methods. The profile refinement techniques are to a large degree responsible for this development. There is hardly any field in crystallography where the Rietveld, or full pattern method has not been tried. The most important recent application is probably quantitative phase analysis.

Profile fitting and pattern decomposition programs, especially using the Two Stage method, allow one to extract individual intensities, thereby determining intensities without knowledge of symmetry or structure of the compound. This opens the way to use powder diffraction data for *ab initio* structure determination by conventional methods, like with Patterson maps, through Fourier calculations or by "direct methods".

Powder diffraction today is widely used in X-ray and neutron diffraction, including the analysis of data collected with synchrotron radiation. Here one important application is their use in high pressure research, for example with diamond anvil cells. In neutron diffraction, besides the study and determination of crystal structures, it is the determination and analysis of magnetic structures, where often complicated magnetic moment configurations with helices for example have to be determined. Texture analysis using neutron diffraction is another growing field of using powder diffraction.

The Rietveld method and the Two Stage method are not in competition. They are two methods with different approaches for the same goal, the analysis of powder diffraction patterns. Each method has its merits and drawbacks. Both require a good mathematical description of the diffraction profile, and powerful least squares routines. The Rietveld method always requires as a *conditio sine qua non* a structural model. The unparalleled power of the Rietveld method is then found in the refinement of rather complicated structures or diffraction patterns with many, and often strongly overlapping peaks, due to low symmetry, large unit cells, etc. Applying the Rietveld, full pattern approach to quantitative phase analysis is perhaps the most promising future application. After a search-match scanning procedure, provided today commercially by the companies marketing diffractometer hardware, the compounds in the specimen are known, and known are also the crystal structures. With this knowledge the structures of both, or even several compounds can be refined and from the scale factors their percentage content in the sample can be calculated.

The Two Stage method finds its power if the crystal structure, e.g. the arrangement of atoms in the unit cell, or magnetic structures are not known. The extracted individual intensities for each reflection allow an analysis by conventional methods, Patterson or Fourier calculations, or even "direct methods". After the structure has been solved, the atomic positions can be refined either by a separate specific program, POWLS for example, or by a full pattern refinement. Many such examples are found in the literature. The Two Stage method is especially useful in problems of non-conventional nature. The use in high pressure research is one such example and one of the most important fields, because the patterns are difficult to analyze in one single formalism.

Contents

Chapter 1

General Considerations

1.1
Introduction

1.1.1
Introduction

X-ray powder diffraction is a powerful non-destructive testing method for determining a range of physical and chemical characteristics of materials. It is widely used in all fields of science and technology. The applications include phase analysis, i.e. the type and quantities of phases present in the sample, the crystallographic unit cell and crystal structure, crystallographic texture, crystalline size, macro-stress and microstrain, and also electron radial distribution functions. Due to the importance and impact on science and technology this technique has been standardized in the European Standard Norms, ESN, (former DIN in Germany) and is documented in the corresponding documents PrEN (WI 138079, WI 138080, WI 138081, WI 138070).

X-ray diffraction results from the interaction between X-rays and electrons of atoms. Depending on the atomic arrangement, interferences between the scattered rays are constructive when the path difference between two diffracted rays differ by an integral number of wavelengths. This selective condition is described by the Bragg equation, also called "Bragg's law":

$$2d_{\mathrm{H}}\sin\Theta_{\mathrm{H}} = n\lambda \tag{1.1}$$

where λ is the wave length, d_{H} the d-spacing and Θ_{H} the Bragg angle, which is half the angle between incident and reflected beam. H describes the Miller indices triplet hkl of each lattice plane. Three sources of radiation are important: X-rays, synchrotron radiation and neutrons. The laws of diffraction, i.e. the interference of diffracted beams holds equally well for all radiations. Electron diffraction is not considered here. Furthermore in this monograph only structural aspects are considered, independent of radiation. It is limited to coherent and elastic scattering.

X-rays: Used in laboratories, where data are collected with sealed X-ray tubes or from rotating anode tubes.

Synchrotron radiation: A beam of charged particles, in general electrons, strongly accelerated in an electric field and deflected in magnetic fields emits besides other electromagnetic radiation a continuous spectrum of X-rays, that is as much as 10^{13} times as brilliant as from sealed X-ray tubes. This radiation is called "synchrotron radiation". The increased brilliance relates to the total energy spectrum.

Monochromatization of the beam typically results in diffraction intensities several orders of magnitude greater than from conventional sources. Synchrotron radiation is used widely today, especially when extreme conditions of the sample environment are required, like high or low temperatures, or high pressures. A further widely used application is the study of kinetics phenomena.

Neutrons: A neutron is an elementary particle with finite mass m and spin 1/2, without electric charge. It carries a magnetic moment and according to de Broglie neutrons behave like waves with wavelength λ and give rise to diffraction:

$$\lambda = h \,/\, (mv) = h \,/\, \sqrt{(2mE)} = 0.286 \,/\, E \tag{1.2}$$

Neutron diffraction is based on nuclear interaction between neutrons and matter on the one hand, and on magnetic interaction with magnetic moments of the atoms due to its magnetic moments. This is the basis for the investigation of magnetic ordering and magnetic structures. The elucidation of magnetic structures is a major application in neutron diffraction (Will 1969b,c, 1972). The (ordered) magnetic moments show sometimes very complicated arrangements of the spins or magnetic moments (Will 1968, 1969c, 1971). This will be discussed.

To obtain a diffraction pattern of a specimen two experimental methods can be used, independent of radiation:

1. The "angular dispersive technique" where the X-rays or neutrons are monochromatic and the pattern is obtained by step-scanning the detector with small increments $\Delta(2\Theta)$. The increments, i.e. the step size may be between 0.02° and 0.001° in 2Θ. The decision for the chosen step size is governed besides instrumental, i.e. mechanical conditions of the diffractometer by the time available to collect the diffraction pattern. Let us assume a range in 2Θ from 5° to 105°, typical for example when using Cu Kα radiation, 1.54 Å, this results in 5 000 counts with $\Delta(2\Theta) = 0.02°$ and requires 1.4 hours. A step size of $\Delta(2\Theta) = 0.001°$ results in 100 000 counts or 28 hours, assuming one second/step.

   Recent developments in instrumentation have led to position sensitive detectors which are used more and more in powder diffraction. Commercially available detectors are in general one-dimensional, which record a large portion of the diffraction pattern simultaneously without moving the detector. For the analysis there is no difference to step-scanning procedures. Two-dimensional detectors are also becoming available, which record the complete Debye-Scherrer ring if used with Debye-Scherrer geometry. They are useful to overcome the problems of preferred orientation.

2. The "energy dispersive technique" where polychromatic X-rays or neutrons are used and the energy of the diffracted X-rays or neutrons is measured at a fixed diffraction angle 2Θ = constant. Energy dispersive technique is especially advantageous in experiments at extreme conditions or for kinetic studies, since the complete pattern is available at all times. In the case of neutrons the energy dispersive technique is gaining importance with the availability of spallation sources (Will et al. 1994; Schäfer et al. 1992b, 1993).

The diffraction patterns look the same in both cases. The difference is seen in the abscissa, where the 2Θ-step scan value is replaced by an energy value E_H. The transformation from 2Θ to energy is simple (Eq. 1.3) The analytical technique, i.e. the analysis of the diffraction patterns discussed here is independent of radiation and diffraction techniques.

$$2d_H \sin\Theta_0 = \lambda_H = hc / E_H = 12.4 \ (\text{keV Å}) / E_H \qquad (1.3)$$

1.1.2
Meaning of the Word "Powder"

The term "powder", as used in powder diffraction, does not strictly correspond to the usual sense in the word in common language (prEN 13925-1, i.e. WI 138079). In powder diffraction the specimen can be a "solid substance divided into very small particles" But it can also be a solid block for example of metal, ceramic, polymer, glass or even a thin film or a liquid. The reason for this is that the important parameters for defining the concept of a powder for a diffraction experiment are the number and size of the individual crystallites that form the specimen, and not their degree of accretion.

An "ideal" powder for a diffraction experiment consists of a large number of small, randomly oriented crystallites (coherently diffracting crystalline domains). If the number is sufficiently large, there are always enough crystallites in any diffracting orientation to give reproducible diffraction patterns. To obtain a precise measurement of the intensity of diffracted rays, the crystallite size must be small, i.e. typically 10 μm or less, depending on the characteristics of the specimen, like absorption, shape etc., and the diffraction geometry.

The analysis by diffraction with different radiations holds equally well for crystalline materials as for amorphous, glassy materials or liquids. This monograph is limited to the analysis of diffraction diagrams collected on crystalline samples. It is limited to coherent and elastic scattering. In one avenue we are concerned with the refinement of crystal structures, which are basically known. In a second avenue the aim is to determine *a priori* unknown crystal structures (in direct competition to single crystal diffraction on, for example, computer controlled four-circle diffractometers). There is of course a gap between powder diffraction data and single crystal diffraction data, but this gap closes more and more as time goes on especially if we refer to inorganic compounds. Crystal structure determinations of organic compounds, especially with "large" molecules, and of course in protein research, where single crystal diffraction is without competition.

1.1.3
Sample and Specimen

We must distinguish between sample and specimen. The term "sample" does not strictly correspond to the general meaning. The sample may be a rather large portion of material, or a very tiny amount. In case of metals or rocks it may be a large, compact piece of material, from which a specimen is taken. When the sample is a rock, it consists in all likelihood of several phases, i.e. minerals, and often with large crystallites. In such cases it is difficult to obtain a "good" sample with a representative distribution of the phases. The sample may consist of several phases, known or unknown, and may also include amorphous material. From the sample a specimen is taken. Depending on the technique and radiation, it may be small or large (in neutron diffraction for example), it may be flat, Bragg-Brentano geometry, or cylindrical, Debye-Scherrer technique. If we are confronted with a multiphase sample, or if we have amorphous contributions which show up as amorphous background in the pattern, the Two Stage method is especially well suited and gives an easy access to separate phases and contributions. This also opens the way to a quantitative analysis.

1.1.4
Why Powder Diffraction?

The analysis of crystalline material and the determination of crystal structures, e.g. the distribution of atoms or ions in crystals, is an important and a well known method since its beginning in the early 1930s. A few examples in the early 1920s crystal structure analyses relied on single crystals with a rapid development in methods, for example Patterson analysis, and technique in the 1930s. The analysis today is highly automated and sophisticated, but it is still based on diffraction experiments with X-rays on single crystals in sizes around 100 μm. The development in instrumentation led to computer controlled four-circle diffractometers, which allow fast measurements on a routine basis. The development in methods of crystal structure analysis with so-called direct methods make the determination of crystal structures, for example of organic material, to a question of hours or a few days. The main application of single crystal structure determination today is found in organic materials and proteins.

So why the analysis of polycrystalline samples, e.g. of powders. Powder diffraction is a very fast and a very versatile analytical method. It is not in competition to single crystal work. Very often single crystals are not available, at least not in the quality or size required for single crystal work. Second, polycrystalline samples can be brought under "extreme" conditions, like high and low temperatures and/or high pressures without too much difficulties. Third, minor impurities are not a serious problem, and finally exposure times can be very short, even in terms of μs. The investigation of kinetics, for example in phase transformations, or of reactions can be studied quite easily and routinely. Last but not least, however the most important feature, the complete diffraction pattern is recorded in a very short time. And if we are using energy-dispersive techniques, the pattern is at each second the complete pattern, a very important property if the sample changes with time, for example in the investigation of kinetic properties.

A further very important application is phase analysis of mixtures of (crystalline) materials. This is probably the fastest growing application, in part because of its importance in industrial analytical laboratories. It will be treated in detail. While in its beginnings in the 1930s the search had to be made manually using cards and books, it moved in the 1950s and 1960s to PC-computers with search-match routines with data stored on diskettes and discs, and finally today it is an almost one-step analysis by the use of full pattern programs.

1.1.5
Difficulties and Limitations of Powder Diffraction

A characteristic feature of powder diffraction is the collapse of the three-dimensional reciprocal space of the individual crystallites on the one-dimensional 2Θ axis. The resulting effects are

i   systematic overlapping of diffraction peaks due to symmetry conditions, for example in cubic space groups;
ii   accidental overlapping because of limited experimental resolution;
iii considerable background difficult to define with accuracy;
iv non-random distribution of the crystallites in the specimen, generally known as preferred orientation.

1.1.6
Fundamentals of Diffraction

Any diffraction experiment is a Fourier transformation from direct or crystal space into reciprocal space yielding intensity data in reciprocal space (Fig. 1.1). Detectors register intensities $I(hkl)$, which are directly proportional to the squares of the crystallographic structure factors $F(hkl)$ (Eq. 1.4): The intensity I is proportional to $|F|^2$. F is a complex quantity.

$$F(hkl) = \sum_j f_j \exp(hx_j + ky_j + lz_j) \tag{1.4}$$

with f_j the form factor or atomic scattering factor of atom j, hkl the Miller indices and xyz the relative atomic positions in the unit cell. The summation j runs over all atoms in one unit cell. Talking of crystalline materials $F(hkl)$ is the Fourier transform of a single unit cell (superimposed on the Fourier transform of the crystal lattice, the reciprocal lattice). Consequently we observe intensities only at the reciprocal lattice points (Fig. 1.2). The intensity being the square of the structure factor leads only to the absolute value of F, $|F|$, while the phase is lost. The phase of the structure factor however is needed for further investigation of the crystal structure. Determination of the phase is the main obstacle in crystallography. After the phase, or sign +/− in centrosymmetric crystals is known we can go back into direct space again by a Fourier transformation, this time by computer, Eq. 1.5:

$$\rho(xyz) = \sum_h \sum_k \sum_l F(hkl)\exp(hx + ky + lz) \tag{1.5}$$

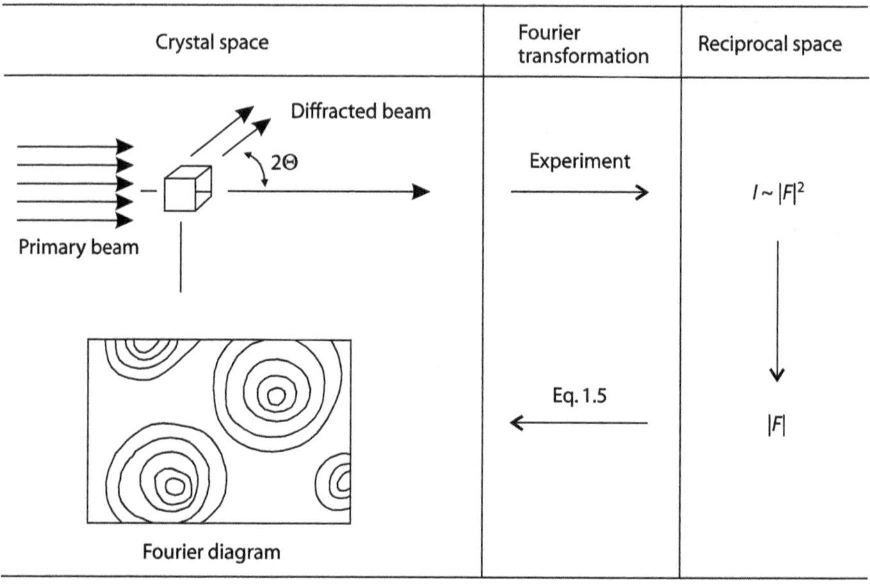

Fig. 1.1. We have two spaces, the direct or crystal space, and the reciprocal space, connected by Fourier-transformation. The experiment itself is a Fourier-transformation

Fig. 1.2. The intensities on the reciprocal lattice points are averaged over spheres of identical d^*-values and superimposed into a one-dimensional diffraction pattern

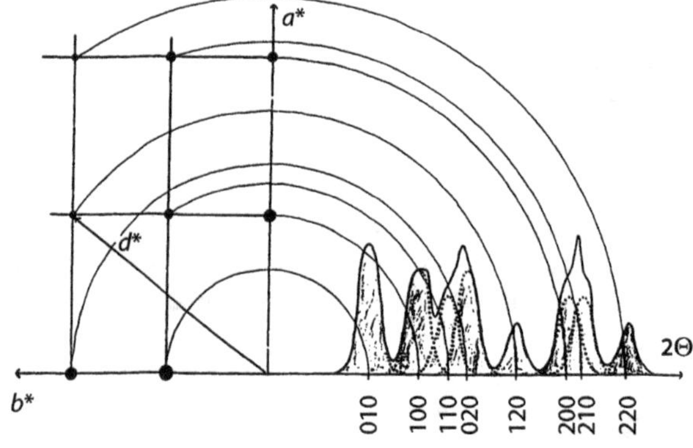

For diffraction on polycrystalline samples the 3-dimensional intensities in reciprocal space are averaged over identical d^*-values and superimposed into a linear, 1-dimensional pattern, as shown in Fig. 1.2. As an example Fig. 1.3 shows the powder diffraction pattern of silicon measured with synchrotron radiation with $\lambda = 1.0021$ Å. Measurements were taken out to $2\Theta = 135°$, corresponding to $d = 0.37$ Å giving 25 well resolved reflections, compared to only 12 reflections when measured with Cu Kα radiation to 160°. The advantage of synchrotron radiation is two-fold: first choosing a short wavelength allows one to measure much further into reciprocal space, with

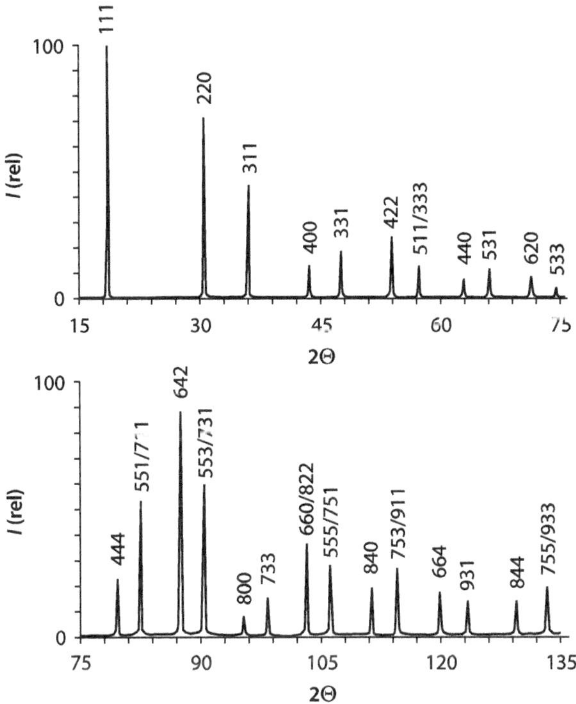

Fig. 1.3. Silicon powder diffraction pattern measured with synchrotron radiation at $\lambda = 1.0021$ Å

still very good intensities and good resolution, second we have single wavelength patterns instead of the α_1/α_2 radiation from sealed X-ray tubes. If required, we can obtain very narrow peaks with good collimators. In the case shown here the resolution was 0.17°.

1.1.7
Analysis of Powder Diffraction Patterns

There are presently two methods used for the analysis of powder diffraction data, the *Rietveld method* or full pattern analysis (Rietveld 1967, 1969) and the *Two Stage method* proposed and developed by Will (Will 1979, 1988, 1989; Will et al. 1983a). For crystal structure refinement both approaches are equivalent and give, and must give, identical results, if applied properly. This has been shown in a comparative study (Will et al. 1990a). The Rietveld method is very fast in its usual applications, which means if it is restricted to the refinement of crystal structures or model structures, which are known at least in principle beforehand. The Rietveld method contains however more potentials as just crystal structure analysis. Such new applications have become known especially in recent years, like "full pattern phase analysis". For general use we have therefore

- The full pattern profile refinement and simultaneous crystal structure refinement in one step. This is the Rietveld method. It is only applicable for refinement.

- The two stage procedure, where in a first step positions and integrated intensities of each reflection are determined by "decomposition" of overlapping diffraction maxima followed by the actual crystal structure calculations. This may be a crystal structure refinement or other applications which require Fourier methods. Even texture investigations for pole figure representations, or structure analysis by direct methods are possible without much difficulties.
- "Pattern decomposition" is in reality "only" the separation of overlapping diffraction maxima. This can be done
  - by full pattern analysis with programs like FULFIT or similar programs (see for example Pawley 1981). This is just one step short of a Rietveld (crystal structure) refinement,
  - by dividing the whole pattern into segments and analyzing each section individually, if necessary by user interaction, like with the program HFIT (Will and Höffner 1992)

Profile fitting is always required, or included in Rietveld's method. This is a least squares fit of a (given and known) profile function to the diffraction pattern by minimizing the function

$$\phi = \sum_i w_i (y_i(\text{obs}) - y_i(\text{calc}))^2 = \text{Minimum} \tag{1.6}$$

with y_i the measured (and calculated) intensities at each step, or channel, in the energy dispersive mode. The summation index i is running over all points in the diffraction pattern, or in some routines running over the section of the pattern selectedfor fitting. A whole diffraction pattern consists typically in the order of 4 000 to 15 000 data points, if we assume for example 40 to 100 point per degree representing 40 to 150 degrees in 2Θ. Those values are very common. This is the range of data if the whole diagram is to be fitted, for example if the program FULFIT or if a Rietveld program is used. If the pattern is segmented into sections, for example commonly used for the analysis of synchrotron diffraction patterns with the fitting program HFIT (Will and Höffner 1992), those numbers are much less. The range of the segment to be fitted is adjustable by the user.

In Eq. 1.6 weights w_i are given to each observation. They are taken from the experimental error margins σ_i, which are assumed to be proportional to the square root of the count rate $y_i(\text{obs})$ following Poisson counting statistics,

$$w_i = 1 / \sigma_i^2 \tag{1.7}$$

The approach to decompose the diffraction patterns before further crystallographic calculations and further analysis is a very flexible and general way. This proves itself the only way if powder diffraction data shall be used for example for structure determinations with direct methods, or in texture analysis.

1.2
Powder Diffraction

1.2.1
History of Powder Diffraction

There is a long and well established tradition of using powder diffraction data to determine crystal structures. We can consider W. L. Bragg as the father of structure analysis. He determined for the first time the crystal structures of NaCl and KCl from single crystals. After the pioneering experiments by M. v. Laue and his co-workers in 1912 W. L. Bragg, at that time a student in Cambridge, England, described in 1913 the structure of rocksalt, NaCl, which was the very first structure to be solved. He was encouraged by Pope, Professor of Chemistry, to try to elucidate the structure of NaCl on the basis of the geometric arrangement of the "spots" in Laue's diffraction patterns. This advice was based on publications by William Barlow in 1883 and 1897. (Barlow 1883, 1897) In those publications he presents pictures of a likely arrangement of atoms and predicts, correctly, the structures of elements and simple binary compounds (see Fig. 165 in Bragg and Bragg 1933, p. 270). Bragg was successful; "the solution of the structure was arrived at by means of Laue photographs and by measurements made with the ionisation spectrometer." He described the structure correctly and he opened thereby the area of crystal structure analysis. To quote Bragg: "We have a new mans of investigating the structure of crystals. Instead of guessing the internal arrangement of the atoms from the outward form assumed by the crystal, we find ourselves able to measure the actual distance from atom to atom. In doing so we seem certain to acquire, indeed we have already acquired, knowledge of great importance to all the sciences and to their applications." In the years following Bragg determined a number of structures, like those of NaCl. KCl, KBr, KJ, CaF_2, Cu_2O, ZnS, pyrite, $NaNO_3$, and some calcites, and also of diamond (Bragg 1913; Bragg and Bragg 1913). These structures were solved mainly by experiments using the "ionisation spectrometer", developed by his father, W. H. Bragg. "The X-ray spectrometer is an instrument designed to make use of the reflection principle." based on this diffractometer Bragg considered the diffraction of X-rays as "reflection from crystal planes" and consequently formulated the "Bragg equation", which is generally used today in crystal structure work rather than the Laue equations.

Powder diffraction followed only two years after Bragg's successful determinations, when Debye and Scherrer developed the Debye-Scherrer technique in 1916: "*Interferenzen an regellos orientierten Teilchen im Röntgenlicht*" (Debye and Scherrer 1916a,b). Debye was at that time professor in Göttingen. They studied and determined the structures of LiF and silicon and they presented basically the complete theory underlying power diffraction patterns, "steigende Quadratsummen geordneter *hkl* triplets" to determine the unit cell dimensions. "Debye's method overcomes the limitations (of larger specimens) and immensely increased the range of crystals which could be examined, since so many substances can only be obtained in a finely divided crystalline form." "The method has, therefore, opened up a new field of the greatest interest and importance." To apply powder data for crystal structural work was slow and also tedious and by no means straight forward. So the earliest examples were inevitably concerned with structures of simple materials, such as iron metal (Hull 1917). Later a limited number of structure determinations were reported, like α- and β-UF_5 (Zachariasen 1949). Such studies were largely based on geometrical considerations and trial-and-error methods. With a number of improvements both methods, Bragg-Brentano and Debye-Scherrer geometry, are still used today, basically unchanged. But the application of powder diffraction for structure determination was limited and for many years drew little attention.

Attention and early developments were concentrated on single crystal methods. Measurements with Bragg's spectrometer were tedious, and it was de Broglie in 1913 who used for the first time the "rotating crystal method". This method was improved by Schiebold in 1919 and further when Weissenberg introduced the Weissenberg camera. Up till about 1924 the success of X-ray analysis has been confined to very simple structures and to solely geometrical arrangement. In order to extent its scope intensities of the reflections had to be known. To calculate intensities it was necessary to know the scattering power of the individual atoms meaning the scattering or f curves. A further step followed when it was realized that a crystal can be considered a three-dimensional lattice, which could be described by Fourier methods. Finally Patterson proposed the Patterson method for the interpretation of Fourier maps. In these years single crystal structure analysis evolved slowly and steadily, still with no real progress and no decisive breakthrough despite the fact, that a number of methods were developed to determine crystal structures from their Fourier and Patterson maps. It was not before the forties and especially after the war that crystal structure analysis from single crystal measurements gained momentum. This can be seen for example in the steadily rising number of publications, and even with new journals dedicated solely to crystal structures, for example Crystal Structure Communications. It can be said that a rapid development of single crystal structure analysis came about because of rapid developments in the field of automated e.g. computer controlled 4-circle diffractometers in combination with faster and more powerful computers. The biggest step forward came with the general availability and ease in using the "direct methods" software for structure solutions. Powder diffraction still was not of significant interest to crystallographers and nobody considered seriously powder diffraction data as a basis for crystal structure determination. Powder diffraction found it place as a useful and very successful method for the identification of materials and for phase analysis.

Looking for a breakthrough in powder diffraction we must focus our attention to neutron diffraction, which was drawing more and more interest, especially in Brookhaven and Oak Ridge, USA, in Saclay, France, in Harwell, England and in Petten, Netherlands, when research rectors became available on a broader basis. Neutron diffraction can be done quite easily with polycrystalline materials and despite the low neutron intensities it was (and still is) very attractive. This enhanced interest was basically a result of the pioneering work by Clifford Shull in 1948 with the determination of the magnetic structure of MnO (Shull and Smart 1949). Neutron research at reactors at that time was concentrated to study magnetic structures, only occasionally crystal structures (Shull et al. 1951). Nevertheless the ordering of magnetic moments occurs in a crystal lattice, whereby the symmetry and structure often changes at the magnetic phase transition. It is therefore not surprising when scientists began wondering how to determine, or at least refine crystal structures from their powder diffraction patterns. This was realized more or less at the same time by two groups, Hugo Rietveld in 1965 at the Reactor Centrum Nederland in Petten, Netherlands, and Georg Will in 1962 at that time working at the Brookhaven National Laboratory in Upton, Long Island, NY, USA. Both were working in neutron diffraction and both had experience in crystal structure analysis.

1.2.2
Programs for the Analysis of Powder Diffraction Data

Powder diffraction has become one of the most active fields in crystallography during the last 15 years. It has become a major source of collecting data with X-rays, neutrons or synchrotron X-rays for crystallographic analyses. Since the times of Bragg and Debye dramatic improvements have been obtained, in hardware as well as in soft ware, meaning programs for the analysis of powder diffraction data. Efforts would have been vain without the dramatic computer advances. The programs with its many version based on Rietveld's ideas of full pattern analysis is the best known example, moving from large scale computers to PCs today. Such programs were made generally available and consequently a wide spread activity has followed to write, to improve and to upgrade existing versions and to adapt programs to one's own requirements. All programs are based on powder diffraction data and are designed for diffraction patterns, normally for X-ray but also for neutron diffraction data.

As a general line on any analysis of a powder diffraction patter the following steps are required:

i   Collection, if possible, of highly resolved powder diffraction patterns
ii Indexing the powder pattern
iii Determination of intensities
iv Moving into one's own problem with specific programs

Smith and Gorter (1991) have published a good compilation of basically all programs available at that time. For details we refer to that compilation. Over 280 programs for the analysis of powder diffraction data have been identified in this compilation, which have been grouped in 21 categories of major types of calculation (see Table 1.1).

Indexing is always the first step in analyzing a powder diffraction pattern. This goes through experimentally available d-spacings, and the single most common programs in powder analysis are based on d-spacings. A first group of programs contains the generation of d-spacings from crystallographic data of known compounds, which allows one to check on possible relationships to known structures. Many programs are provided which give a graphical display based on crystallographic data either from d-I (d-spacing – intensity) data sets using the PDF data file or from atomic positions in the structure. If these steps are successful phase identification is possible and in general straight forward.

There are always diffraction patterns, where indexing on this simple approach is not possible. Automatic indexing is a line of programming, which dates back to the 1960s with more sophisticated routines starting in the 1970s. Here the most impor-

Table 1.1. Subdivision of the programs for the a analysis of powder diffraction data in categories of major types of calculation

Crystallographic databases	7
Analysis packages	17
Instrument control and data processing	34
d generation	4
Graphical $d - I$ display	8
Phase identification, including search-match	17
Automatic indexing	15
Refinement/indexing	12
Refinement/error analysis	9
Metric analysis	18
Pattern generation	22
Profile fitting – decomposition	24
Profile fitting – full pattern	7
Deconvolution	6
Crystallinity/strain/stress/texture	9
Rietveld structure refinement	25
Quantitative analysis	18
Structure determination (from powder data) "Direct methods"	2
Structure display	26
Small-angle scattering	8
Miscellaneous programs	12

tant factor which controls the success of indexing is the accuracy of the data. If preliminary indexing, often only on a very limited and well resolved number of peaks in the forward region, has been reached a refinement procedure can be started, generally in an iterative mode including more and more reflections. One of the earliest and best known program is probably the APPLEMAN program written in the late 1950s (Appleman and Evans 1973). All programs to follow are derivatives of this routine. However in the last years, since about 2000, significant progress has been made in developing new and more straight forward indexing algorithms (see Sect. 1.2.3).

There are complete program systems available today which incorporate most of the computer programs necessary for analysis of powder diffraction data including Rietveld formalisms. One such system is GSAS, derived and compiled by Larsen and Von Dreele (1994). This system is "public domain". More recently a similar program system has been developed by Coelho and his co-workers (Cheary and Coelho 1992; Coelho et al. 1997). This system is marketed today by the Bruker/Siemens Company.

Many applications require individual intensities. This is not just the case in the Two Stage method, but for many other methods, like Patterson mapping. With the availability of accurate digitized diffraction patterns, peak analysis, cluster separation and profile fitting is a popular option with many programs available. Twenty-four are given in the list. The fitting considers each peak or peak cluster as a single fitting. A step further is the "full pattern profile fitting", where a number of program are available. All such programs do not require information on the structure. They give reliable information on lattice parameters, *FWHM*, and (integrated) intensities, without requiring detailed information on the material. *FWHM* for example is the basis to be used for particle size analysis or strain/stress analysis. The intensities are given for each reflection, including overlapping ones in clusters, which are used with the usual single crystal analysis programs including also direct methods programs. This is the basic idea of the Two Stage approach. It was used in all *a priori* crystal structure determinations from powder diffraction data found in the literature.

Crystallite size is commonly calculated by the Scherrer method by analyzing the breadth of all peaks in the pattern:

$$t(hkl) = 0.9\lambda(B_0^2(2\Theta) - B_{ST}^2(2\Theta))^{1/2}\cos\Theta \tag{1.8}$$

with t the particle size, B_0 the profile width of the specimen and B_{ST} the profile width of a standard sample, like silicon. This simple formula assumes equal dimensions in the particles. For more detailed analysis the line broadening for different directions may be considered, like $(h00)$ and $(00l)$ to determine cylindrical dimensions. Macrostrain can be measured by measuring peak shifts relative to an unstrained sample. Both, crystallite size and strain properties can be analyzed by using a Williamson-Hall plot. Another category is the variance method, which is an accurate technique for studying crystal imperfections. These applications however are beyond the scope of this monograph.

1.2.3
Crystal Structures from Powder Diffraction Data

The earliest examples of structures solved from powder diffraction data were of simple materials, such as iron metal (Hull 1917). Much later a limited number of structure determinations were reported by Zachariasen, like α and β UF_5 (Zachariasen 1949). Such studies were based on geometrical considerations and trial-and-error methods. For many years the field of crystal structure determination was basically monopolized by single crystal methods, in hardware as well as in methods and software.

Detailed and precise information on a crystal structure can be obtained only from three-dimensional diffraction data. For many years they could be obtained only from diffraction experiments with single crystals. There are however a number of reasons why single crystal diffraction is not possible, for example under extreme environmental conditions, like low or high temperatures, high pressure etc., or quite general if single crystals of the compound to be studied are just not available, or difficult to make. Powder diffraction experiments are an alternative. Two efforts, by Will in 1962, and by Rietveld in 1966 brought crystal structure analysis, at the beginning only of refinement to powder diffraction. At this time single crystal structure analyses were considered to be far superior to analyses from powder diffraction data. This has changed in the last decades, and today in many cases analyses based on powder diffraction data are very well comparable to single crystal data. This is especially true if the compounds in question are within certain limits determined by the size of the unit cell, symmetry and complexity. But it must be stated with some despair that there seems to be some exaggeration in the race for solving the largest structure from powder data, even claiming protein structures.

Any crystal structure analysis, be it with single crystals or with polycrystalline samples, begins with a structural model, even when using direct methods. This model is then refined on the basis of experimental data. There are a number of methods developed for single crystal analyses, and those methods hold equally well for powder diffraction data, if individual intensities are available. Refinement is the second step after a model has been established. Even the Rietveld method is nothing else than a complex minimization procedure built on external knowledge of a preconceived model. This can in the course of the refinement only slightly be modified. The starting parameters must be reasonably close to the final values, and more over the sequence into which the different parameters are being refined needs to be carefully selected. The refinement needs experience. Rietveld approach cannot be taken as an active tool for structure determination.

Indexing

In order to solve a crystal structure from powder diffraction data as the first step the pattern must be indexed. About 50–90% of all structure determination attempts fail because of failing indexing. Is powder indexing a solved problem? Not really. Scientists devoted their time very early to this problem and a number of programs are available to help in indexing a diffraction pattern. See for example Visser (1969); DICVOL91 by Louer (Louer and Louer 1972; Boultif and Louer 1991); TREOR by Werner (Werner 1964, 1976; Werner et al. 1985). Those algorithms have served the scientific community for decades; and they are adequate for most indexing problems, especially for volumes less than 1 000 A^3. Since the 1990s apparently no significant progress has been made in developing new and especially more straight forward and more reliable indexing algorithms. This situation has changed in the last years, when a number of programs have been published, for example by Coelho (2003). Nearly all these mature indexing programs have their roots in the 1970s.

In practice indexing programs will never provide a single and unambiguous solution. Indexing is not a black-box approach; solid crystallographic background is a requirement. Use of several programs can maximize the possibility to find, and also to identify a correct solution. Taking advantage of different algorithms gives some idea of the range of different solutions and helps if identical solutions or derivative cells can be identified. Common requirements for successful indexing are sample purity and line-position data of the highest possible quality. The most popular indexing programs today probably are ITO, TREOR, and DICVOL.

ITO is a deductive search program by zone-indexing in index space; it performs best when given 30 to 40 accurately measured powder lines. It is optimized for low symmetry systems (orthorhombic downwards); high-symmetry lattices may get reported wrongly in an orthorhombic setting. TREOR is a semi-exhaustive, heuristic search method in index space; it requires about 25 accurately measured powder lines. It is effective for searches down to triclinic symmetry. DICVOL again is an exhaustive search program in parameter space by successive dichotomy; it requires around 20 accurately measured powder lines and is well suited for high symmetry down to monoclinic, impurity lines however are not tolerated. In all cases the success depends on accurate peak positions.

Recently more advances in powder indexing has been provided by Siemens/Bruker in a system called the SVD-index method and LP-Search. SVD-indexing (SVD = Single Value Decomposition) (Coelho 2003) operates on d-values extracted from reasonable quality powder diffraction data. It can operate as an iterative process with hkl's assigned, or by a Monte Carlo approach searching the parameter space. It is not an exhaustive method. The programs ITO, TREOR and DICVOL have been incorporated into this package. Another approach, in the same package, is LP-Search (Coelho 2003), a Monte Carlo based whole powder pattern decomposition program. It is independent of d-spacing extraction and is therefore suited for indexing of poor quality powder data. No d-values are required as input data.

After indexing the pattern the actual analysis for determining crystal structures form powder diffraction data has to begin with profile fitting and profile refinement techniques. For this the peak shape must be known. The better the reflections can be described the better are the results afterwards. One serious problem is that the profiles in general change with diffraction angle. A great number of crystal structures, even complex crystal structures have been studied in recent years. Even if "Rietveld refinement" appears in the title, in reality a structural model has been deduced before by "conventional" methods, and this model was then refined by a "Rietveld" program.

Determination of a crystal structure from powder diffraction data follows the following steps (see Table 1.2):

i   Collection of a highly resolved powder diffraction pattern
ii Indexing the powder pattern, determination of the unit cell and space group
iii Determination of intensities, resulting in a list of hkl and I (Miller indices and intensities). For solving unknown structures the intensities may be of limited accuracy
iv Structure solution using for example Patterson methods, direct methods or any other general approach
v   Structure refinement by "Rietveld" or POWLS

1.2.4
Pattern Decomposition for Accurate Intensities

The determination of unknown crystal structures needs intensities and structure factors $F(hkl)$. Those values can easily be extracted if well resolved peaks in powder diffraction patterns are found. A better, and more sophisticated way is by computer based pattern decomposition. Pattern decomposition, or synonymous profile fitting is the first step in a "Two Stage" approach, as it is advocated by Will. It opens powder diffraction with its individual hkl data to any crystallographic problem, especially solving unknown crystal structures by standard crystallographic methods. This can be done by full pattern fitting. Rietveld refinement then follows as a consequent ensuing step. Besides analyzing the full pattern as a whole, there is sometimes a need to treat the diffraction pattern in sections on an interactive individually operator controlled decomposition. Table 1.3 lists a classification of such methods.

Table 1.2. The procedure for crystal structure determination from powder data

Sample preparation

Data collection

Determine peak positions

PDF – ICDD

Indexing

CDF-search

Unit cell, symmetry and space group

Integrated intensities

Derive structural model
(Patterson, direct methods, etc.)

Fourier methods

Refinement by Rietveld or POWLS

Table 1.3. Classification of methods

	Profile fitting method	**Whole-pattern decomposition**	**Rietveld refinement method**
Aim of analysis	Pattern decomposition	Pattern decomposition and refinement of unit cell parameters	Structure refinement
Range of analysis	Local peaks or peak clusters	Whole pattern	Whole pattern
Peak position	Independent parameters	Function of unit cell parameters	Function of unit cell parameters
Integrated intensities (profile area)	Independent parameters	Independent parameters	Function of structural parameters
Initial parameters required to start refinement	Nil	Approximate unit cell parameters	Initial structural parameters (structural model)

The peak overlapping is an intrinsic problem in analyzing powder diffraction data. The method, which decomposes the powder pattern into individual Bragg components without reference to a structural model is called as pattern decomposition method. This is in contrast to the Rietveld method or full pattern refinement, which requires a structural model. There are many cases where an unsolved structure does not give sufficient information on atomic parameters to start with the refinement of the structure. Example are the *ab initio* structure determination from powder diffraction data with conventional methods, like calculating Patterson maps, applying heavy atom or isomorphous replacement methods, or even by direct methods. Especially with direct methods it is of the utmost importance to have as many and as accurate unique intensities as possible. All these methods have been developed originally on the basis of single crystal data. But there is no reason and no intrinsic problem to translate these methods to powder diffraction data, provided integrated intensities can be extracted from the diffraction patterns with sufficient accuracy. Overlapping intensities are unavoidable in powder diffraction, and more over the range of measurable intensities in a powder pattern is much smaller than in single crystal diffractometry. Both, overlapping reflections and the limited 2Θ range complicates the process of crystal structure determination. These difficulties can be tackled by advanced computation. For such cases crystallographers have developed methods to arrive at a correct model for the crystal structure to be used for further refinement.

Whole Pattern Fitting

Originally simple approaches, like integrating peaks by planimeter, or just taking the peak heights of the reflections were used. Today a number of programs for the determination of accurate intensities from powder diffraction data have been developed. These programs yield accurate, and even very accurate intensities by whole-pattern fitting based on least squares procedures. The first step in all cases is the indexing of the reflections and the subsequent calculation of the lattice parameters. Very reliable programs are available to perform this task. The next step is the decomposition of the intensities. Whole pattern fitting programs require as input only the wavelength and the lattice parameters, which may be approximate values and in general are included in the least squares decomposition routines. Since integrated intensities are needed, profile fitting methods are the only choice. Consequently a reliable mathematical description of the profiles is required, and this is the most difficult parameter.

One of the first attempts dates back to a paper by Pawley (1980, 1981). This early program was written with neutron diffraction data in mind and consequently using the more simple Gaussian description of the diffraction peaks. The program was original intended to determine and refine lattice parameters, but giving also integrated intensities of each reflection. This early program had several shortcomings. For example good intensity values were obtained accurately only when the peak distance to the next neighbour was larger than the halfwidth of the peak. A profoundly improved version of "whole pattern fitting with a least squares procedure" was provided by Jansen J. et al. (1992). In this program suite all pertinent parameters are included in a standard refinement – background, cell dimensions, intensities and peak shape. In view of many shortcomings in powder diffraction patterns the refinement technique used by Jansen, Peschar and Schenk (1992) tackled the refinement problem in two steps, executed interactively until an acceptable convergence has been reached. In the first step intensities are refined together with the background. In the second step all other parameters such as peak shape, zero point and cell dimensions are refined, keeping the parameters of the first step constant, and vice versa.

Another program suite is FULFIT by Jansen E. et al. (1988) and Will et al. (1990a), successfully used in several applications. It is a full pattern refinement without knowledge about the crystal structure. The program needs as starting values only the instrumental parameters R, S, and T describing the variation of *FWHM* according to Cagliotti et al. (1958) and approximate values for the lattice parameters. It comes very close to the method of full pattern structure refinement (Rietveld's method) refining the whole pattern except atomic parameters of the (unknown) structure. It uses the input files from POWLS (Will 1979) for Laue and space group symmetries, which makes it extremely user-friendly. In addition to the lattice parameters the peak positions 2Θ and peak heights for each reflection are varied and refined. Integrated intensities are calculated using peak heights and *FWHM*. FULFIT generates a list of *hkl* values, *d*-spacing and intensities. This list is determined by the space group symmetry excluding forbidden reflections. It may be modified by the user. Besides the integrated intensities the output contains also the standard deviations and correlations, i.e. the so-called variance-covariance matrix for a correct weighting scheme to be used in the ensuing structure refinement. Refinement of the wavelength (sometimes advisable in neutron diffraction) and zero point $2\Theta_0$ is possible. Figure 1.4 and Table 1.4 give an example for Li_2CuO_2 measured with neutrons. Included in Table 1.4 are for the section $2\Theta = 50$ to $70°$ *hkl*, lattice spacing *d*, integrated intensities *I*, standard deviations σ and percentage correlations ϱ_x to the subsequent peaks. The *R*-index R_{PF} was 2.9%.

Figure 1.5 gives a further example of FULFIT, where a diffraction pattern of olivine, $(Mg,Fe)_2SiO_4$, collected with synchrotron radiation, $\lambda = 1.74050$ Å. Olivine crystallizes orthorhombic, Pbnm, with $a = 4.7631$ Å, $b = 10.2270$ Å and $c = 5.9946$ Å. This gives many peaks which are especially at higher angles heavily overlapped. It was no problem to decompose the pattern and to extract integrated intensities. They were used to calculate three-dimensional electron density maps (Will 1989). Since this mineral was

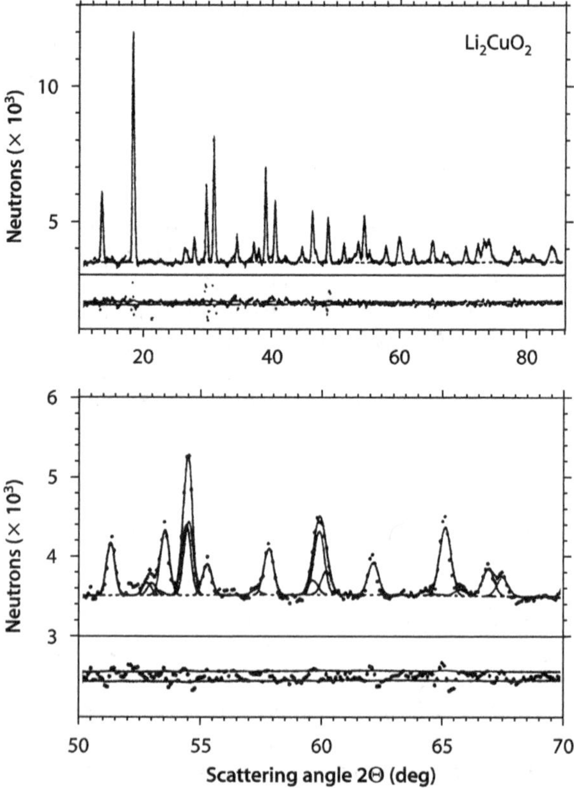

Fig. 1.4. Observed (*dots*) and calculated (*line*) diffraction patterns of orthorhombic Li_2CuO_2. The profile analysis used FULFIT. A selected part $2\Theta = 50$ to $70°$ is shown on an enlarged scale, where the individual reflection profiles are clearly visible

available also as single crystals it was also measured with a four circle diffractometer, so a comparison could be made giving full agreement.

A comprehensive summary of the whole-pattern decomposition method isgiven by Toraya in *The Rigaku Journal* (Toraya 1989). There he refers primarilyto his own program, WPPF (Toraya 1986). The method of whole-powder pattern decomposition decomposes the powder pattern, e.g. the whole pattern in contrast to interactive decomposition programs, where selected regions of the pattern are analyzed, into individual Bragg components without reference to a structural model. Summarizing the following programs, all for step-scanned angle dispersive diffraction patterns, shall be mentioned: ALLHKL by Pawley (1981), FULFIT by Jansen E. et al. (1988) and WPPF by Toraya (1986). Pattern decomposition of patterns collect with energy dispersive or time-of-flight method are discussed in Sect. 3.11.

Interactive Pattern Decomposition

Full pattern decomposition fails, if a diffraction pattern is not well resolved, or has several components, etc. In such cases interactive profile fitting methods are a good

Table 1.4. Output of a full-pattern profile fit from FULFIT on Li_2CuO_2

hkl	d (Å)	I	σ	ρ_1 ($\times$ 100)	ρ_2	ρ_3	ρ_4	ρ_5	ρ_6
220	1.127	2787	293	0					
222	1.096	1072	380	−63	24	0			
321	1.092	3577	432	−46	0				
125	1.087	1965	333	0					
026	1.056	2034	288	0					
305	1.024	424	276	0					
314/217	1.013	4563	304	−2	0				
109	1.004	688	278	0					
208	0.998	1854	286	−4	0				
019	0.980	1349	283	0					
031	0.949	655	332	−51	0				
127	0.946	2481	340	0					
321	0.924	2320	642	−88	2	−1	0		
130	0.923	1272	639	−2	1	0			
226	0.915	2502	404	−67	16	−11	6	−1	0
316/033	0.912	2325	416	−28	19	−11	2	0	
028	0.908	2697	516	−79	47	−11	0		
132	0.906	1475	638	−71	18	0			
307	0.903	1824	441	−29	0				
402	0.899	808	294	0					

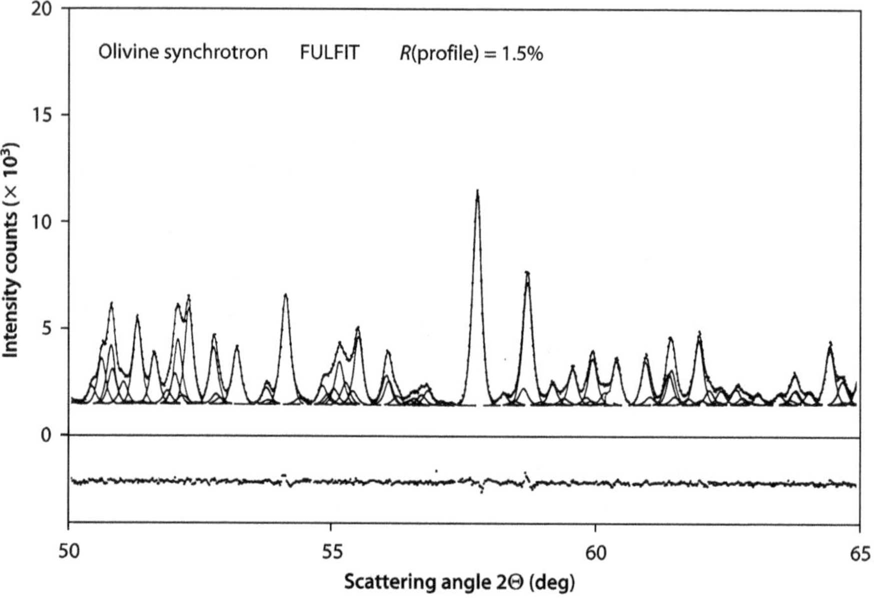

Fig. 1.5. Section of an olivine diffraction pattern as an example for a full-pattern-profile analysis with FULFIT. Shown is the section 2Θ = 50 to 65°. The differences between observed and calculated data points are plotted at the bottom. The profile R-index for this section is 1.5%

alternative. Difficulties of this kind are found for example in patterns collected with synchrotron radiation in the energy dispersive mode under extreme conditions like pressure or temperature. They cannot be refined by Rietveld methods, and can not even be analyzed by full pattern decomposition. Here interactive programs are useful and needed. One program is HFIT provided by Höffner (Höffner and Will 1991; Will and Höffner 1992), originally designed for the analysis of synchrotron diffraction patterns collected in diamond anvil cells under high pressures. Another program is PROFAN (Merz et al. 1990), designed for neutron diffraction patterns in the analysis of magnetic structures. Both programs are designed for interactive, computer aided fitting, where each peak or peak cluster is treated as a single fitting. The programs are interactive for on-line-operation on a graphic display by setting the cursor on the proper position. The operator can make selections of his choice. For example by pressing ENTER an additional peak can be entered in the fitting calculation, or a peak can be deleted. Programs of this type are of particular interest in the analysis of unknown materials, of low or uncertain symmetry crystal data, or of multiphase diffraction patterns.

For example in high pressure research the patterns have besides the peaks from the specimen also other peaks for example from the pressure marker, the gasket, escape peaks, etc. or even from several specimen phases, like when high and low pressure phases are present and disturb the pattern. At the start and as a preliminary step the pattern is divided into sections, each containing just one peak or a peak cluster selected with the cursor for on-line analysis. An additional difficulty results from a possible sudden change in the background, which may occur in energy dispersive setup if the energy goes through an absorption edge. Such an example is shown in Fig. 3.19 in Sect. 3.3.2. Figure 1.6. shows an example when using laboratory X-rays with $K\alpha_1/K\alpha_2$ splitting for a diffraction pattern of NiNdC2.

The other example PROFAN operates on similar grounds. Again preselected peaks or peak clusters from neutron, X-ray or synchrotron diffraction patterns are analyzed individually. Figure 1.7 gives an example for such an interactive analysis. The profile fitting is based on a nonlinear weighted least squares routine (Hamilton 1964a), as is used also in POWLS, by minimizing the sum of weighted squared differences between observed and calculated counting rates. Several commonly used profile shape functions are implemented. In neutron diffraction patterns sometimes of complicated magnetic structures are observed, stemming from helical configurations (Kockelmann et al. 1992a,b; Kockelmann 1995). For such cases a specific version of POWLS has been written, IC-POWLS, which can handle, and refine almost all magnetic moment configurations confronted with in neutron diffraction research (Kockelmann et al. 1995). All programs require no information on the structure.

The general line of analysis of a powder diffraction pattern for a compound with an unknown structure should proceed with the followings steps:

i   Collection of powder diffraction patterns
ii Determination of approximate unit cell parameters, sometimes based on isomorphous compounds, or by the many indexing programs
iii Determination of integrated intensities by pattern decomposition, either with full pattern decomposition programs, or by interactive pattern decomposition
iv Moving into one's own problem with specific programs

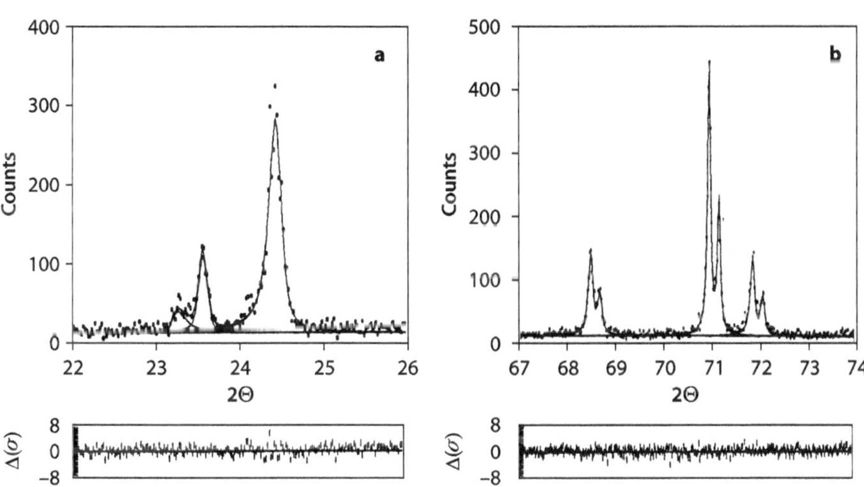

Fig. 1.6. X-ray diffraction pattern of $NiNdC_2$; **a** between $2\Theta = 22$ and $26°$; **b** between $2\Theta = 67$ and $74°$. The $K\alpha_1/K\alpha_2$ splitting can be seen

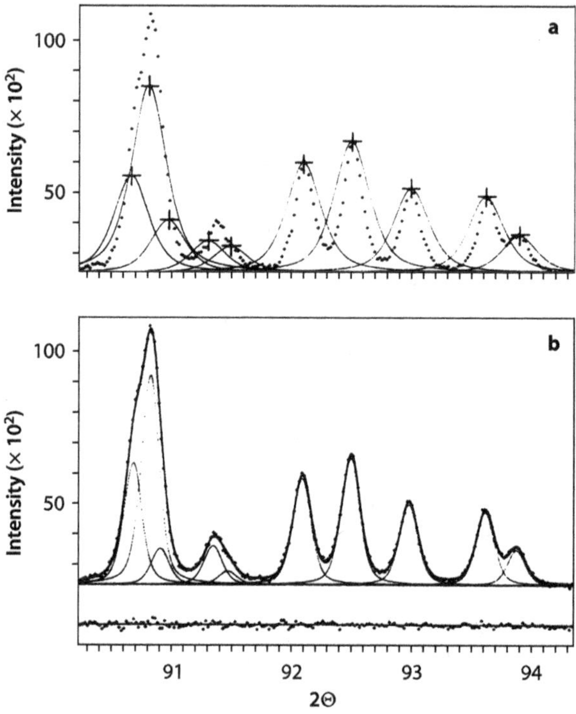

Fig. 1.7. Interactive profile refinement of a section of quartz reflections measured with synchrotron radiation (*data points*). **a** Input of starting values by selecting peak maxima with the cursor (cross markers clicked with the mouse). The corresponding starting profiles are plotted. **b** Final result of the profile fitting performed with 10 Pseudo-Voigt functions. The R-index is 2.0%

1.2.5
Profile Shape Functions

A number of functions have been developed and are used in Rietveld programs as well as in profile fitting programs for the Two Stage method. The first description of the profile shape of laboratory X-ray diffraction peaks comes from Parrish and co-workers (Parrish et al. 1976). They used seven Lorentzian functions for one α_1/α_2 peak, also as a function of diffraction angle. This description is cumbersome and tedious and is not used any more. Today four functions are widely used: The Gaussian G, the Lorentzian L, the Pseudo-Voigt function V, and the Pearson VII function P. They are compiled in Table 1.5. Sometimes the real Voigt function, a convolution of G and L is suggested. The Voigt and Pseudo-Voigt functions combine a Gaussian and a Lorentzian function. All functions, including the description by Parrish, are used for pattern decomposition (Two Stage method). For full pattern refinements the description by Parrish is not suitable.

Using the real Voigt function has not found much enthusiasm. In one paper Suortti et al. (1979) have analyzed powder diffraction profiles of Ni measured with X-rays and neutrons using a Voigt function fit. In the neutron case the instrumental factors are dominant, while the X-ray profiles are determined by the size and strain of the particles. In conclusion this study reveals however, that the true Voigt function, besides extensive mathematical programming, is not useful for structure analytical purposes.

Table 1.5. Profile shape functions commonly used in profile fitting programs

Gaussian	$G = I_0 \exp\left(-\ln2\left(\dfrac{2\Theta - 2\Theta_0}{\omega}\right)^2\right)$
Lorentzian	$L = I_0\left(1 + \left(\dfrac{2\Theta - 2\Theta_0}{\omega}\right)^2\right)^{-n} \quad n = 1;1.5;2$
Pseudo-Voigt	$V = \eta L + (1-\eta)G \quad (0 \le \eta \le 1)$
Pearson VII	$P = I_0\left(1 + \left(\dfrac{2\Theta - 2\Theta_0}{ma^2}\right)^2\right)^{-m}$

Parameter: $2\Theta_0$ = peak position; I_0 = peak intensity; ω = FWHM/2; m = shape parameter.

It can be successfully used only for a profile which can be completely resolved, that means for simple structures with only a few and well resolved reflections. To quote the authors: "in other words the Voigtian refinement can be used only in cases where it is not needed".

Of these profile shape functions the Pseudo-Voigt and the Pearson VII functions produce the best fit between observed and calculated peak profiles, since both incorporate a variable shape parameter. The Pearson VII function has some features to recommend because it allows a continuous variation of the shape parameter (m) from pure Lorentzian through the intermediate and modified Lorentzians to pure Gaussian. The Pseudo-Voigt function on the other hand applies a variable proportion of pure Lorentzian to pure Gaussian by the parameter η. There are several modifications of the Lorentzian type function with exponents -2 or -1.5 (see Table 1.5). The full width at half maximum, $FWHM$, is normally and very well described by the well known formula published by Cagliotti et al. (1958).

Which function is the best to describe the observed intensity profiles? It was believed by some researchers that differences between the peak shape model and reality may produce a distortion of the refinement results. Therefore the influence of the shape functions on the results has been studied in some detail by Young and Wiles (1982), and by Hill and Howard (1987), the latter for neutron diffraction data. In neutron diffraction the peak shape is predominant Gaussian. Nevertheless Hill and Howard used a Pseudo-Voigt function and studied the influence of the Lorentzian contribution described by the parameter η. Three compounds were investigated and compared to single crystal data: TiO_2, PbO and β-PbO_2. They refined the data with three fractional Lorentzian characters, $\eta = 0$ (pure Gaussian), $\eta = 0.112$ and $\eta = 0.08 + 0.0004 \times 2\Theta$, i.e. in the last case including a variation with diffraction angle. The results are in all cases basically the same and identical with single crystal data. This very detailed study has shown that the structural parameters are hardly affected by the peak shape function. It also casts some light on the sometimes exaggerated influence of η on the refinement, at least as far as it concerns structural and unit cell parameters.

The other compilation and comparison of different profile shape functions comes from Young and Wiles (1982) with X-ray diffraction data. This study also shows, that positional parameters are generally not affected beyond standard deviations by the choice of the function used. Thermal parameters however often are.

Factors which do influence the line shape are crystallite size and microstrain contributions. The crystallite size broadening produces L-shaped intrinsic tails in the peak profile, while microstrain produces G-shaped contributions. *FWHM* values are significantly affected with such samples. Since those data are collected however in different connections, they are of little importance for structure refinements, where the aim is to determine precise structural parameters, the common goal of almost all Rietveld refinements.

Thermal parameters on the other hand are affected by the choice of the description of the profiles. These quantities are to be used with extreme care, if at all. They hardly represent the thermal movement of the atoms in the structure. The Debye-Waller factors obtained in profile fittings and full pattern refinements are all together not reliable quantities for the description of thermal motions of atoms in the unit cell.

Serious sources of errors are the correct handling of the background, especially if it is included in the refinement. And thermal diffuse scattering (TDS) is another obstacle which is hard to model, even in single crystal studies. And this is one of the most serious reasons why thermal parameters, i.e. Debye-Waller factors, hardly describe thermal movements of atoms correctly.

In summary unit cell edge and atomic positional parameters are largely unaffected by the assumed model for the peak shape and are close to the values obtained from single crystal diffraction. Also it has been shown that agreement indices and the structural parameters are in most cases the same, at least within the quoted e.s.d. values whether the peak shapes are fitted to a Pseudo-Voigt or a Pearson VII function.

Criteria of Fit

At the end of a refinement it is necessary to check whether the results are meaningful, and whether they meet certain standard criteria. This holds both for single peaks and for full pattern refinements. The overall best criteria for the fit are without doubt difference plots between observed and calculated data.

a   Criteria for single resolved reflections
  - y_i(obs) – y_i(calc) difference plot
  - maximum deviation at any point in the difference plot
  - goodness of fit, R-index for one (or several) isolated individual reflection
  - variance range, observed vs. calculated
b   Criteria for the whole pattern in a Rietveld analysis
  - y_i(obs) – y_i(calc) difference plot
  - maximum deviation at any point in the difference plot
  - index R_{wp} for the profile fit
  - index R_B (B for Bragg) for the agreement of integrated peak values. It must be kept in mind that the values I(obs) are calculated on the basis of the calculated ones and are therefore not directly comparable to single crystal R-indices. Struc-

tural parameters and their standard deviations (if possible in comparison to similar single crystal results)

It is customary to calculate "goodness of fit" indicators, known as R-indices. They are compiled in Table 1.6. Powder diffraction data are not directly comparable to single crystal data, but the R-indices used are close to the single crystal structure R-value. Only R_p and R_wp shown in Table 1.6 are of real value.

Of the several R-indices which are used to check on the quality of the fit and refinement R_wp is statistically the most meaningful indicator of the overall fit since the numerator is the residual that is minimized in the least squares procedure. Any R-index presented is meaningful only if the background has been subtracted. This deserves a word of caution. The influence of background on R-values is significant. It has been discussed in detail by Jansen E. et al. (1994a). The contribution from a high background enters the denumerator and will thus greatly affect the R-indices. It implies better fitting results than actually valid.

R_B, the so-called Bragg R-index, implies a comparison of integrated intensities similar to single crystal refinements. In a Rietveld refinement there are no real integrated intensities and R_B is based on somewhat fictitious observed intensities. It is calculated by allocating the actual observed (step scanned) intensities y_i(obs) to Bragg intensities on the basis of the calculated intensities on a "share holder" basis. Nevertheless it is a useful indicator for judging the results.

All R-indices are greatly affected, if the crystallites are not ideally "imperfect", e.g. if they are affected by crystallite size and/or microstrain. Crystallite-size shows up in line broadening and can be recognized easily in the y_i(obs) – y_i(calc) difference plots. It produces intrinsic contributions to the Lorentz profile. Microstrain on the other side produces Gaussian shaped profiles, and again this can be recognized in the y_i(obs) – y_i(calc) difference plots. Very general the R-indices as quantities for fit and accuracy must be taken with care. One might assign in such cases some significance to η in the Pseudo-Voigt formula, particularly to its angular dependence. This should be a reason to include η as a parameter in the refinement.

Table 1.6. Definition of R-indices used in Rietveld analyses

$$R_F = \frac{\sum \left| (I_K(\mathrm{obs}))^{1/2} - (I_K(\mathrm{calc}))^{1/2} \right|}{\sum (I_K(\mathrm{obs}))^{1/2}}$$

$$R_B = \frac{\sum \left| (I_K(\mathrm{obs})) - (I_K(\mathrm{calc})) \right|}{\sum I_K(\mathrm{obs})} \quad (= K_I)$$

$$R_p = \frac{\sum \left| y_i(\mathrm{obs}) - (1/c)y_i(\mathrm{calc}) \right|}{\sum y_i(\mathrm{obs})}$$

$$R_{wp} = \left[\frac{\sum w_i (y_i(\mathrm{obs}) - (1/c)y_i(\mathrm{calc}))^2}{\sum w_i (y_i(\mathrm{obs}))^2} \right]^{1/2}$$

1.2.6
Early *Ab Initio* Crystal Structure Determinations from Powder Diffraction Data by Pattern Decomposition

Scientists working in neutron diffraction were the pioneers to study crystal structures from powder diffraction data. They were facing the problems of lacking single crystals sufficiently large for single crystal studies, and in other cases they needed lattice parameters and atomic positions of the materials studied for their investigations of magnetic structures. One such example, probably the first one to refine a crystal structure from powder data was $MnSO_4$ by Will et al. (1965). $MnSO_4$ is isostructural with $NiSO_4$. In the course of a neutron diffraction study the agreement between observed and calculated nuclear intensities from the neutron diffraction powder pattern was poor when using atomic parameters from isomorphous compounds. Both X-ray and neutron diffraction data were then collected and the crystal structure refined using intensities from reflections integrated with planimeter. For refinement they used the least squares program POWLS, where groups of overlapping reflections, $\Sigma i\, jF_i^2$ were taken as one observation (Will 1979). A R-index of 3.3% was reached.

Another notable example of complete crystal structure determinations from powder diffraction data, again both from X-ray and neutron measurements, is that of Ca_3UO_6 and Sr_3UO_6 by Rietveld (1966). Rietveld began this investigation with the generally used procedure: He measured the compounds with a high resolution Guinier film camera to determine symmetry (monoclinic) and unit cell (Ba_3UO_6 was also measured, but the Guinier data did not allow reliable indexing and this compound was not studied further). The structural investigation was continued by calculating 3D-Patterson maps, which lead to the positions of uranium and calcium. Those were used for phasing Fourier maps yielding in rather rudimentary pictures the complete structure. Difficulties were encountered however in the least squares refinements because of severe peak overlaps. For this reason Rietveld wrote a especially adapted least squares program for the Electrologica computer, available in Petten. In this program he calculated structure factors for overlapping groups of reflections, $\Sigma r\, (jF^2(\text{obs}) - \Sigma r\, jF^2(\text{calc}))^2$ and considered each group as one observation in the least squares refinement. This was necessary, because the overlaps in composite reflection groups could not be separated and reduced to its individual structure factors, so the composite sums $\Sigma r\, jF^2$ were taken. This is the same procedure as taken independently by Will on $MnSO_4$ in Brookhaven with the program POWLS. For the calcium compound 45 intensities comprising 100 independent reflections were measured from the X-ray diffractogram. Both compounds were also measured with neutrons and their structures refined with similar success. The R-values reached were around 5 to 6%.

This procedure to determine or refine crystal structures was not very attractive. Rietveld therefore put his attention to separate composite peaks with the help of a computer program. His first publication in this direction was concerned (only) with profile fitting in neutron diffraction patterns by a least squares formalism. Since the diffraction profiles in neutron diffraction can be described mathematically by Gaussian functions, this method could be handled fairly straight forward with the computer power available at that time. This lead to integrated individual intensities for further crystal structure determinations and ensuing refinements with conventional least squares programs (Rietveld 1966). The examples were Ca_3UO_6 and Sr_3UO_6 with 29 positional parameters (Loopstra and Rietveld 1969). Another such example is $Ti_2Nb_{10}O_{29}$ with 41 positional parameters (Von Dreele and Cheetham 1974). The actual full pattern refinement, what we call today Rietveld analysis came several years later (Rietveld 1969). Rietveld's idea of "pattern decomposition" with the help of computer programs was taken up first by Australian scientists from the laboratory where Rietveld came from.

Before the full pattern least squares refinement, Rietveld refinement, of (known) crystal structures became widely known, a number of publications appeared following the early idea of Rietveld to extract intensities by profile decomposition, i.e. profile refinement. The first mention of profile fitting in order to extract reliable intensities based on Rietveld's publications is found in several publications by Taylor and Wilson in 1974/75, in compounds with heavy elements: β-tungsten hexachloride, uranium hexachloride and uranium tetrabromide (Taylor and Wilson 1974a–c), or by Levy, Taylor and Wilson in 1975 on uranium(III)triiodide, molybdenum hexafluoride and some other fluorides (Levy et al. 1975a,b). In all these cases data were collected by neutron diffraction in Australia. For the actual refinement of those crystal structures the conventional single crystal least squares program ORFLS by Busing and Levy was used. Taylor et al. (1986) also published a paper using X-ray diffraction, when they studied and determined the crystal structure of a zeolite sample, H^+-ZSM5. In this specific case a special computer program was written for a point-by-point decomposition of the profile for each (hkl). The structure itself was refined using the single crystal least squares program SHELX.

Except from a few examples in the early days the general way to determine intensities from powder diffraction patterns was by simply taking the peak heights as the intensity values. The first publication where an unknown and more complex crystal structure was solved purely from powder diffraction data appeared in 1977 (Berg and Werner 1977). The compound was $(NH_4)_4[(MoO_2)_4O_3](C_4H_3O_5)_2 \cdot H_2O$, a rather complex structure. Single crystals could not be obtained and it is frustrating when a new compound with unknown structure cannot be obtained as single crystals (of reasonable size). For simple compounds often structures can be deduced from similarities with known structures. But when the compound is more complex the situation is difficult. Data were collected from a microcrystalline sample with a focusing Guinier-Hägg camera with monochromatic Cu $K\alpha_1$ radiation. Exposure time was two hours. The film was evaluated using the film scanner developed and available in the laboratory in Stockholm. The peak position allowed indexing and the determination of the unit cell: monoclinic, space group C2, $a = 14.572(2)$ Å, $b = 10.114(2)$ Å, $c = 11.461(2)$ Å, $\beta = 121.45°$. Integrated intensities were provided by the film scanner program system, which provides step-scan intensities and also calculates integrated intensities (Malmros and Werner 1973). The step size was 0.01586 degrees. With the intensities the crystal structure was determined using general crystallographic methods.

They approached the problem in three steps: (*i*) first to determine from peak positions unit cell, symmetry and space group, then (*ii*) to determine intensities and with them (*iii*) to follow the conventional routine of structure determination: Patterson maps to derive a structural model, Fourier and difference Fourier maps to locate the atoms in the unit cell and finally to refine the structure by least squares methods, here with a Rietveld program. Basically the procedure is the "Two Stage" approach.

The structure determination was started with three-dimensional Patterson maps calculated with 120 resolved integrated intensities. All heavy-atom peaks could be seen and the positions for molybdenum could be assigned. A first (conventional ORFLS) least squares refinement with the molybdenum positions only ended with a rather poor R-index of only 0.30. No oxygen position could be derived. Then a Rietveld profile analysis and refinement program, rewritten for Guinier data by Malmros and Thomas (1977) was applied. Within the observed two-theta range 505 intensities and consequently structure factors could be derived, compared to 129 values before. For a first Fourier calculation 250 data for $(\sin\Theta) / \lambda > 0.336$ were used. By successive Fourier and difference Fourier maps and ensuing least squares calculations all atoms, except hydrogen, could be localized and the structure fully determined. The final least squares profile refinement with 505 reflections and 51 parameters varied ended with $R = 0.11$, a very acceptable result considering the complexness of the structure and film data as the basis for the intensities.

In these examples peak shape was not a problem or a parameter. Extending the analysis and attempting peak decomposition for X-ray diffraction data posed serious difficulties in the beginning. Peak shape in X-ray diffraction pattern do not confirm to a description by a Gaussian function. Second there is the α_1/α_2 doublet, and more serious a variation of peak shape with scattering angle. When Khattak and Cox (1977) approached Rietveld refinement of X-ray diffraction data they used the single line β-radiation. For their study they chose $La_{0.75}Sr_{0.25}CrO_3$, a compound studied previously with neutrons. This allowed a direct comparison between neutron and X-ray results. In addition it gave valuable information on the profiles to be used. The main question in this paper was the peak shape and the problem of pattern fitting without reference to a structural model. It was a very notable, so not the first attempt. For example Huang and Parrish (1975) described the measured, observed profiles with a rather formidable composite of seven Lorentzians involving 21 parameters. It gave an excellent fit to their diffractometer $K\alpha$ profiles, including the handicap of $K\alpha_1$ and $K\alpha_2$. Khattak and Cox tried several functions with the single Cu $K\beta$ radiation and they found that the peak shapes were described reasonably well by a modified Lorentz-type function. Comparing their results with the ones from previous neutron data demonstrated this a workable functions. (Later and today the Pseudo-Voigt function is preferred, a composite of a Lorentzian and a Gaussian function.)

Another early example is the structure determination of α-$CrPO_4$ from powder synchrotron X-ray data (Attfield et al. 1986). With the generally very good resolution of synchrotron diffraction patterns 68 well resolved peaks were obtained. They were used, after corrections for multiplicity and Lorentz factor, to generate a Patterson map, which revealed the two positions of the two heavy atoms, chromium and phosphorous. They were used to start phasing the data for the Fourier maps. With several Fourier maps the remaining atoms could be localized. The structure was finally refined using a modified version of the standard Rietveld program using a Pseudo-Voigt function to describe the peak shapes.

After the efforts by Berg and Werner in 1977 there was little progress in structure determination from powder diffraction data until the mid-1980s. One of the first structure determinations from individually determined intensities without prior knowledge of the structure is that of $ZrKH(PO_4)_2$ by Clearfield et al. (1984). Their data were collected by X-ray diffraction. They started by the rather rudimentary method to determine integrated intensities by cutting out areas under the peaks and weighing them. This yielded 40 to 50 usable pieces of data. The positions of the zirconium atom could be determined from a Patterson map, and the remaining atoms were found by Fourier methods. Since the positions of the atoms were now fairly well known the structure was finally refined with the Rietveld technique using the 10 to 80° data set. The final R indicators were $R_F = 0.036$ and $R_{wp} = 0.145$.

Daniel Louer with his group began to use powder diffraction data for the determination of crystal structures since the late 1980s. He reported several successful determinations, all of low symmetry. Examples are $Nd(OH)_2NO_3 \cdot H_2O$ (Louer and Louer 1987), $Zr(OH)_2(NO_3)_2 \cdot 4.7\,H_2O$ (Bernard et al. 1991a), $Cd_5(OH)_8(NO_3)_2 \cdot 2\,H_2O$ (Bernard et al. 1991b), or $LiB_2O_3(OH) \cdot H_2O$ (Louer et al. 1992).

$Nd(OH)_2NO_3 \cdot H_2O$ crystallizes monoclinic, space group C2/m (Louer and Louer 1987). Indexing is always the crucial stage in a structural analysis. It was accomplished with the successive dichotomy method using the program DICVOL (Louer and Vargas 1982). By an ensuing pattern decomposition using a Pseudo-Voigt function for the profile shape 71 integrated intensities with unambiguous index assignment were obtained. Beginning with a three-dimensional Patterson map the position of neodymium could be assigned, followed by successive Fourier maps where the eight remaining atoms could be located. The full structure was refined with the Rietveld program DBW2.9 (Young and Wiles 1982) with 19 atomic coordinates, 9 isotropic temperature factors, a zero-point parameter, 4 unit cell parameters and 3 halfwidth parameters. The weighted-profile R-index R_{wp} was 0.17.

$Zr(OH)_2(NO_3)_2 \cdot 4.7\,H_2O$ crystallizes triclinic, pace group P$\bar{1}$ (Bernard et al. 1991a). After indexing 136 integrated intensities could be extracted by pattern decomposition. They were used to calculate a Patterson map, from which approximate heavy atom coordinates were derived. The remaining atoms were located by successive three-dimensional Fourier maps. The final structure was again refined by the Rietveld method ending with a weighted-profile R-index R_{wp} of 0.145. In the calculation 65 positional parameters were refined from 896 reflections in the angular range $2\Theta = 12$ to $92°$.

$Cd_5(OH)_8(NO_3)_2 \cdot 2\,H_2O$ is a complex structure with a large unit cell, $a = 18.931$ Å, $b = 6.850$ Å, $c = 5.931$ Å and $\beta = 94.85°$, and monoclinic symmetry, space group C2/m resulting in a heavily crowded powder diffraction pattern. The data were collected on a Siemens D500 powder diffractometer operating with strictly monochromatic copper $K\alpha_1$ radiation. The diffraction pattern was decomposed into its Bragg reflections with the Siemens DIFFRAC-AT software using a Pseudo-Voigt function. The shape factor was $\eta = 0.98$ indicating a strong Lorentzian character of the line shape. From the precise powder data the automatic indexing program DICVOL gave the lattice parameters.

The structure was determined from a 3-dimensional Patterson map using 96 unambiguously indexed reflections from the powder data set. Twenty-three unobserved, but space group allowed reflections with small, but non-zero intensity were included in the calculation of the Patterson map. The approximate atomic positions were then used for a least squares structure refinement by the Rietveld method using the program DBW3.2S. The refinement used 444 powder reflections. It included 40 parameters, 19 atomic coordinates, 5 isotropic temperature factors, the scale factor, 4 cell parameters, a zero-point parameter, an asymmetry parameter and 6 parameters to define the functional dependence of the background. The refinement resulted in a only partly satisfactory R-index value of $R_B = 0.083$ (B for Bragg), a profile value $R_p = 0.093$ and $R_{wp} = 0.116$. The final difference plot shown in the publication is fairly good, but it must be realized that the measured data beyond $2\Theta > 60°$ disappear in the background and can hardly be made responsible for a contribution to the refinement.

A similar structure of the same series, $Cd_3(OH)_5(NO_3)$ has been determined directly by conventional standard methods from powder diffraction data with monochromatic X-rays (Plevert et al. 1989). The cell dimensions of this compound are $a = 3.4203$ Å, $b = 10.0292$ Å, $c = 11.0295$ Å, space group Pmmn with $Z = 2$. From pattern fitting integrated intensities, line shape and line width were obtained. To derive a structural model a Patterson diagram was calculated using 65 unambiguously indexed reflections. This gave the positions of the cadmium atoms. The remaining atoms were located by the interpretation of Fourier difference maps. Finally the structure was refined by whole-pattern refinement converging to $R_F = 0.049$, $R_B = 0.059$ and $R_p = 0.061$.

A further example of structure determinations using pattern decomposition and the Rietveld method on synchrotron X-ray and neutron powder data, e.g. with the Two Stage method is presented by Lehmann et al. (1987). The compounds studied were $Al_2Y_4O_9$ and I_2O_4. The diffraction patterns were first analyzed by profile decomposition using the program ALLHKL (Pawley 1981), which gave individual structure amplitudes. This was followed by applying standard crystallographic methods. For the solution of the structure of I_2O_4 first a Patterson map was calculated with the extracted structure amplitudes. This yielded the two I atoms. Followed by repeated Fourier and difference Fourier calculations, doing also repeatedly preliminary Rietveld refinements to allocate all atoms, gave finally the locations of the oxygen atoms. In $Al_2Y_4O_9$ the Y atoms were located by direct methods with the program MULTAN77 (Main et al. 1977), again followed by Fourier and difference Fourier calculations.

In both cases structural models were derived by conventional methods, Patterson methods and direct methods, and then refined using Rietveld methods Especially the combination with Fourier and difference Fourier calculations improved the electron density maps and revealed missing atoms. The final structures were refined with full pattern Rietveld programs. Using the X-ray and neutron data sets separately gave rather high R-values, ranging from 7.7 to 12.2%. Combining both data sets did improve the R-values considerably, $R = 3.1$ to 6.9%, and also resulted in acceptable standard deviations of the atomic positions.

A very notable example of an *ab initio* structure determination has been published by McCusker (1988) for Sigma-2, a new clathrasil phase. High resolution data were collected with synchrotron radiation and the decomposition of the diffraction pattern was obtained by full pattern refinement using the program ALLHKL (Pawley 1981), yielding 258 reflections. This publication is especially noteworthy, since the structure was solved by direct methods. A detailed discussion is given in Sect. 1.2.7.

One more structure solved *ab initio* by direct methods is $LiB_2O_3(OH) \cdot H_2O$ (Louer et al. 1992). A first data set was collected using Bragg-Brentano geometry for indexing the powder pattern by the successive dichotomy method. The final data were collected in the laboratory with Cu Kα radiation using a curved position-sensitive detector from INEL. Integrated intensities were extracted by means of a fitting procedure using a full pattern decomposition program FULLPROF written by Rodriguez-Carajal. It is derived from the Rietveld program DBW3.2S and works very similar to FULFIT by Jansen E. et al. (1988). From 269 structure factor values the direct method program

MULTAN was used giving first approximate atomic positions. By successive Fourier and difference Fourier calculations the complete structure could be derived and finally refined with a Rietveld program. The refinement involved 23 atomic coordinates, 9 isotropic temperature parameters, scale factor and zero point correction, 3 cell parameters and 2 parameters describing the angular variation of the mixing factor η in the Pseudo-Voigt expression. The final R-index was $R_{wp} = 0.04$.

Finally to mention is the paper by Estermann et al. (1992) "*Ab initio structure determination from severely overlapping powder diffraction data*". They investigated SAPO-40, a aluminophosphate-based molecular sieve with small amounts of Si in the tetrahedral framework sites. With $a = 22.045$ Å, $b = 13.699$ Å, and $c = 7.120$ Å, space group Pmmm, and 21 atoms to be located this is a very large unit cell and more than a standard problem. Integrated intensities for all reflections with 2Θ less than $70°$ were extracted from the powder pattern with the Program EXTRACT, developed at the ETH Zürich. When attempts with several direct method programs failed a new method was brought into use, the so called "fast iterative Patterson squaring (FIPS)", which involves an iterative procedure of treating points in the Patterson maps by squaring. This gave finally the correct structure.

These examples are all two-stage procedures. They start with the determination of (integrated) intensities, usually by pattern decomposition, as a basis for the structure determination, in general with conventional Patterson and Fourier methods, in some cases with Hauptmann's direct methods. As the pertinent point refinement always requires a model structure. If such a model has been found the final structure can be refined by Rietveld, full pattern, or other least squares calculations.

For completeness there is at least one publication, where the structure was solved directly from the X-ray powder diffraction data with the Rietveld procedure without decomposition. This was synthetic hercynite, $FeAl_2O_4$, a mineral crystallizing with the spinel structure, which could be used as input for a model structure (Hill 1984). The structure has only three positional parameters. A flexible profile shape of the Pseudo-Voigt type was used with 33.7 to 99.8% Lorentzian character as 2Θ increased. The structure refinements were undertaken with the full-profile Rietveld type program DBW3.2 (Rietveld 1969; Wlles and Young 1981).

1.2.7
Ab Initio Structure Determination by Direct Methods in Powder Diffraction

It is the ultimate goal of crystallographers to overcome the "phase problem" and to solve crystal structures directly from the experimental data. The term "direct methods" indicates those methods which try to derive the phases of the structure factors, which are lost in the experiment, directly from the observed amplitudes through mathematical relationships. A breakthrough came, when Karle and Hauptmann presented in a number of publications such a way of solving crystal structures directly from the measured intensities. The method was developed by Karle and Hauptmann in the 1950s (Karle and Hauptmann 1950; Hauptmann and Karle 1956). For this achievement Hauptmann was rewarded the Nobel Prize in 1964. The two quantities phase and amplitude of the diffracted wave are independent quantities. By statistical considerations it is possible to derive a relation between them. This is based on two important properties: (*i*) the electron density is everywhere positive, and (*ii*) the structure is composed of discrete atoms. Programs are generally available today. A detailed description of this method can be found in Giacovazzo (1980, 1992, 1996).

"Direct method" programs do not give a complete and unbiased structure. They lead to a first approximate solution constituting one or several initial structural models for the atomic distributions in the unit cell. The correct structure must be selected by the operator and refined by least squares methods. Direct methods are routinely used for the determination of structures primarily of organic compounds measured with single crystals. For using it with powder diffraction data there is a fundamental difference: Data from single crystal measurements lead to one, or several structural proposals, which can be tested and refined with the usual least squares programs. When using powder diffraction patterns first a reasonable large number of integrated intensities is required, which are extracted by profile fitting. Even if these limited data sets give basically correct structures, the refinements must be done with full pattern refinement programs. So we are talking here about a two stage process. A few examples solving crystal structures from powder diffraction data are the structure determinations of Sigma-2 by McCusker (1988), of $LiSbWO_6$ by Le Bail et al. (1988) and of $C_{10}N_{16}N_6S$ by Cernik et al. (1991), which will be discussed later.

Care is advised in using the term "*ab initio*", which is not identical with "direct method" solutions. It is occasionally found in the literature pretending a structure solution directly from the intensities or structure factors. This is misleading. "*Ab initio*" means in a literal translation without any previous knowledge concerning the structure. In that sense any crystal structure determination is "*ab initio*". Publications which carry "*ab initio*" in their title try to imply a direct method approach, while actually they have deduced model structures by conventional methods, mostly by Patterson methods.

Direct Structure Solution by Triplet Phase Invariants

It is obvious that it must be possible to apply G. Hauptmann's direct method also for powder diffraction data. However the successful application of this method to powder diffraction data is still a challenge and it can not be described as fully acceptable and successful. The problems encountered are far from trivial. Even if we assume that the indexing and space group problems have been solved there are still serious problems we are facing: (*i*) overlapping diffraction peaks, accidental and more serious systematic, (*ii*) the background which is often difficult to define unambiguously leading to erroneous intensity values, and (*iii*) preferred orientation. After successful pattern decomposition the method begins to reduce intensities to E-moduli. One difficulty arises already here, since peaks, which are heavily overlapped, are also highly correlated. Even good decomposition programs often provide negative, and thus wrong intensities. Any error in the estimate of the structure factor moduli will weaken the efficiency of the method, since incorrect moduli will lead to wrong phases and thus to structure models not acceptable. The use of synchrotron radiation reduces these problems, but we can safely assume that experiments with synchrotrons are the exception rather than the rule. Giacovazzo has put much energy on this problem and developed a number of programs, especially SIRPOW92. In a lead article (Giacovazzo 1996) he lists 41 structures which he has solved from powder diffraction data, a few other examples can be found in the literature.

Direct methods work well when the ratio of independent intensities to the number of atoms to be found is sufficiently high, as a rule larger than seven, conditions found easily in single crystal studies of organic compounds. In practice the structure is first solved by phasing a limited amount of structure factors, in single crystal work up to 500. This number is hard to arrive from powder diffraction patterns. Also because of the unavoidable errors in the phasing procedure and owing to the effects of series truncation the complete structure is usually not evident from the first E-map. Ultimately interpretation of these data is heavily dependent upon experience and intuition, and this is especially true when based on powder diffraction data with often only a limited number of reflections available. Thus successive refinement steps with full pattern refinement programs must follow including Fourier and difference Fourier mapping cycles.

In summary we find, that the great efficiency of direct methods in single crystal work can not be translated directly to powder work, at least not yet. The risk of failure is high, since the limited uncertainty in the structure factor amplitudes is still the main obstacle. Significant improvements can be expected if decomposition is more reliable, especially if the resolution is improved.

Direct Structure Solution by Simulated Annealing

Another approach was taken up recently by Coelho with a method he calls "simulated annealing" (Coelho 2000). The concept of "simulated annealing" is based on the concept that nature tends towards minimum energy. Thereby he derives model structures by a Monte Carlo approach. It may be remembered that a "Monte Carlo" method was suggested in 1960 at an IUCr satellite conference in Glasgow on "Computing Methods" by Niggli/ETH Zürich together with V. Vand (Vand and Niggli 1961). The idea was to obtain atomic coordinates or consecutive sets of random numbers from a random-number generator program by placing atoms randomly in the unit cell and put the model on a refinement cycle for obtaining an initial crystal structure model. Random number generator programs became just available in 1960. Niggli's idea was based purely on a statistical distribution of the atoms in the unit cell without restrictions. Since it required extensive computer time and since at that time computers neither had the required speed nor memory, Niggli's method was not followed further. Today computer power and speed have progressed enormously, so that some time-consuming approaches have become more practicable.

Coelho's method is more advanced. It is not just "trial and error" approach, it is based on scientific considerations and limitations. The method offers considerable success since it does not start with a completely random and arbitrary structure proposals. "Simulated annealing" uses sound physical and above all crystallographic considerations, like known bond length and atomic radii.

The method facilitates a solution of structures from the starting point using only space group and lattice parameters. The significant idea is the inclusion of electrostatic potential penalty functions in a nonlinear least squares Rietveld refinement procedure. In a thermodynamic (macroscopic) process of annealing, materials crystallize over time on slow cooling. If this is translated into crystallography atoms tend to arrange in a minimum-energy configuration and this should lead to an arrangement of atoms resembling the structure observed in nature. To simulate this process is called "simulated annealing". At the start the atoms in a unit cell are randomly thrown into three dimensional space and are allowed to arrive in an energy minimum. In time they should arrange in a manner similar to observed structures. This comprises iterative numerical procedures designed to seek a global minimum. In the paper by Coelho several examples are given with iterations ranging from 1 000 to 5 000. Obviously this method can only work today with modern computers which are large and fast enough. But again, the method presents models which have to be refined afterwards. It is to be seen how successful this new method will be and how it will be accepted by the crystallographic community.

The structure solution of a simulated annealing cycle is performed in several steps (Coelho 2000):

i   Find local minimum of the objective function
ii Decrease the temperature
iii Randomize the atomic coordinates

Repeating such cycles resembles a random walk through parameter space. This method fits directly to powder diffraction data, since it refines the structural proposals with Rietveld methods. An additional very important point is the introduction of a potential-energy function, which describe atoms in real space. A bond-length penalty function in terms of known bond length is included. The method has been successfully applied to several known structures, as listed in Coelho's paper: $AlVO_4$, $K_2HCr_2AsO_{10}$ and $[Co(NH_3)_5 CO_3)]NO_3 \cdot H_2O$.

1.2.8
Crystal Structures Determined by Direct Methods

"*Ab initio*" crystal structure determination using the formalism of direct methods by "triplet phase invariants" according to Hauptmann and Karle is a routine procedure today, if used with single crystal data. A number of programs are available, like SHELX (Sheldrick 1976), MULTAN (Main 1985), or XTAL (Stewart and Hall 1988). There are two programs SIR88 (Burla et al. 1989), and SIRPOW91 (Cascarano et al. 1992) and SIRPOW92 (Giacovazzo 1996) written specifically for using with powder diffraction data. Giacovazzo and his group have concentrated on this problem and developed a number of programs, especially SIRPOW92. In a lead article (Giacovazzo 1996) lists 41 structures which he has solved.

Several publications can be found, where intensity data were derived from powder diffraction patterns and then used with one of the above programs for the solution of the crystal structure. All these publications work with two stages, first to determine integrated intensities, then to determine the structure and in general as the final step to refine the structure by full pattern refinement using the full powder diffraction pattern.

Powder diffraction patterns in general suffer from overlapping reflections making the extraction of individual intensities difficult and often unreliable. With the advent of widely available synchrotron radiation it has become a possible and more feasible approach to this problem: Synchrotron sources provide high intensity X-rays and high resolution, which allow a much less ambiguous decomposition of the powder diffraction patterns than using sealed X-ray tubes. Overlapping reflections are not eliminated, but they are minimized. Once reliable integrated intensity values have been calculated, a single crystal structure determination procedure by direct methods can follow. But we must keep in mind, the direct-method programs do not provide an unbiased solution. Ultimately interpretation of these data is heavily dependent upon experience and intuition of the scientist in front of his computer screen, and this is especially true when based on powder diffraction data with often only a limited number of reflections available. Despite these shortcomings the method is gaining more and more interest, and a few crystal structures solved from powder diffraction data are for example Sigma-2 by McCusker (1988), $(C_{10}N_{16}N_6S)$ by Cernik et al. (1991), and $LiSbWO_6$ by Le Bail et al. (1988).

McCusker (1988) reports such an *ab initio* structure determination of Sigma-2, a new clathrasil phase from a synchrotron powder diffraction pattern. Data were collected at the NSLS in Brookhaven with truly monochromatic 1.5468 Å radiation. Back ground, which in general is low in synchrotron experiments, was subtracted, peak positions and the 2Θ values were input to the auto-indexing program TREOR (Werner et al. 1985). With the symmetry information and unit cell dimensions available the diagram was then input in the whole pattern fitting program ALLHKL (Pawley 1981) yielding 258 reflections. However only 66 had E-values ≤ 1.0. They were used in the direct-method single crystal package XTAL (Stewart and Hall 1988). There are four Si and seven O atoms in the asymmetric unit. Nine of the eleven framework positions appeared in the direct method solution. Eight were used to generate a Fourier map, which clearly showed the remaining three oxygen positions. Finally a whole pattern refinement was followed, which converged with $R = 0.10$.

Cernik and co-workers (1991) studied $(C_{10}N_{16}N_6S)$ with data from a high resolution synchrotron experiment as an example for a small organic molecule. The data were auto-indexed using again the program TREOR (Werner at al. 1985). Integrated intensities were obtained by pattern decomposition giving 561 reflections, of which 204 were rejected because they were statistically unobserved. Several direct method programs failed to solve the whole structure. They could locate only three to four atoms, including sulfur, but not enough to solve the structure. The structure could finally be solved taking the top 17 peaks from the output of the direct method program SIR (Burla et al. 1989), which is specifically written for powder diffraction patterns. The final solution was found from Fourier and difference Fourier maps. The structure was finally refined with a Rietveld full pattern program. The final R-index was 1.9%.

Another successful example is the structure determination of $LiSbWO_6$ by Le Bail et al. (1988). In this case the Hauptmann's direct method formalism was applied, a truly *ab initio* structure determination. The compound crystallizes in the columbite structure. The structure was solved by a two stage process: First the authors extracted individual intensities from the powder pattern by profile fitting, then they indexed the pattern with DICVOL (Louer and Vargas 1982) and determined unit cell and symmetry. As the second step intensities were converted into structure factors and used as input for the SHELX program (Sheldrick 1976). Initial positional parameters of tungsten and antimony were obtained from direct methods applied on 307 reflections. Three oxygen positions were then located from a Fourier difference map. At this stage the Rietveld method e.g. full pattern refinement was used, which is the proper choice in order to handle overlapping reflections. The final R-index was 2.1%.

1.2.9
Preferred Orientation in Powder Diffraction

Powder diffraction is based on a fully random distribution of crystallites of equal size. Any deviation from a random distribution will affect more or less the measured intensities in the diffraction pattern. These deviations are meant by "preferred orientation", e.g. a term for a non-random distribution of crystallites in the specimen. It must not be confused with texture, which has a much larger degree of crystallite distribution and has to be treated differently. Preferred orientation is always a problem when analyzing powder diffraction patterns for correct crystal structure values. This is a serious drawback especially when the intensities are the basis for structure solutions with direct methods, which are based on E-values derived from the intensities. Preferred orientation is one of the major sources responsible for failure when applying direct methods in powder diffraction.

Preferred orientation has been realized already by Rietveld in his first program, Eq. 1.9a (Rietveld 1969), and it has also been considered by Will in his program POWLS Eq. 1.9b (Will et al. 1983a). Rietveld used (rather large) cylindrical sample holder for his neutron diffraction experiments with (elongated) crystallites oriented along the cylinder axis. Will, Parrish and Huang (Will et al. 1983a) used the Bragg-Brentano geometry with the crystallites, elongated or plate-like, in the specimen plane. Considering the different diffraction geometries the following formulas have been suggested, and used, to correct for these effects:

$$I_{\text{corr}} = I_{\text{obs}}\exp(-G\phi^2) \tag{1.9a}$$

$$I_{\text{corr}} = I_{\text{obs}}\exp(-G(\pi/2 - \phi^2)) \tag{1.9b}$$

G is a correction parameter treated as a variable in the refinements. ϕ is the acute angle between the scattering vector (hkl) and a vector (HKL) defined by the operator as the "preferred orientation" vector. These formulas are simple and they work in general exceedingly well reducing the R-index significantly. Preferred orientation in its general meaning of non-random distribution is observed to a small degree in practically all specimens used in powder diffraction (remember specimens, not samples!). If the degree of preferred orientation is higher, like in minerals exhibiting cleavage properties, like calcite, mica, etc. a more advanced formula has been proposed by Dollase (1986) for specimens with considerable deviation from randomness of the crystallites (Eq. 1.9c). This formula is based on a mathematical treatment discussed by March (1932).

$$I_{\text{corr}} = I_{\text{obs}}(G^2\cos^2\phi + \sin^2\phi / G)^{-3/2} \tag{1.9c}$$

Preferred orientation is encountered especially in flat specimens using Bragg-Brentano geometry, where the active volume of the specimen is small. The effect can be minimized by spinning the specimen around the normal to the specimen plane or by loose packing of the powder (see Sect. 3.5.2).

The effect of preferred orientation can be a serious problem in the application of powder diffraction for conventional phase analysis based on the so-called Hanawalt index files based on the three (in modern programs five) largest peaks in the pattern. It may easily lead even to failure of the analysis. Here progress is expected, and partially is already present when using full pattern refinements of the multicomponent pattern. On one hand, deviations in peak intensities are absorbed by considering the complete structure patterns, on the other hand corrections for preferred orientation can easily be included in the programs.

1.2.10
Line Broadening: Crystallite Size, Strain and Stress

Powder diffraction has almost unlimited possibilities. Diversified and wide spread applications can be observed in the continuing number of specific programs written by scientists with quite different objectives in mind. The key to most applications is the analysis of the profiles. The key word is *microstructure*: Polycrystalline materials invariably contain imperfections that modify the intensity of the Bragg reflections, and more seriously also the peak shape. This departure from an ideal structure is known as *microstructure*. The study of structural imperfections by means of powder diffraction, mostly by X-ray diffraction, is known as *Line Profile Analysis*. The main deviations observed in the profiles come from very small particles and from strain and stress in the sample. Since a powder diffraction pattern contains the complete information about the sample: structure, crystallinity, strain, stress, and so on, it is possible to study those effects from powder diffraction data. As a general rule the contribution of crystallite size is approximated by a Lorentzian function of a width proportional to $1/\cos\Theta$. Strain broadening is covered by the Gaussian, which has a width proportional to $\tan\Theta$.

When the Rietveld method was extended to X-ray diffraction this has led in the beginning to difficulties, because there was no general and easy way to use functions to describe the observed profiles. For well behaved samples, e.g. of proper particle size and free of strain, stress or stacking faults the Pseudo-Voigt function gives a good representation of the profile and it is used today in most programs for Rietveld refinements and also in the Two Stage method. It is easy to handle and easy to put into the software. It remains however an approximation. The true function would be the real Voigt function, a convolution of a Gaussian and a Lorentzian function rather than an addition. A true Voigt function is needed therefore.

Microstructure affects the breadth and shape of the diffraction line profile. It may also introduce a displacement from its ideal position. Line breadth and displacement therefore has to be taken into account for a successful analysis. This requires extensive profile fitting, which, if taken properly, will give the wanted information. The study of microstructure in a more or less rudimentary form is as old as X-ray powder diffraction itself. For example the determination of particle size is treated already by the so-called Scherrer equation developed in the course of Scherrer's Ph.D. thesis in Göttingen in 1918.

$$D_{\text{Scherrer}} = K\lambda / \beta\cos\Theta \tag{1.10}$$

with β the integral breadth of the profile, and K a constant, for example $K = 0.9$ for crystallites with cubic, i.e. equal shape. For more detailed determination of the crystallite sizes in different directions, for example elongated, the profiles of several reflections have to be analyzed preferable in the three directions by analyzing the reflections $h00$, $0k0$, $0l0$. The Scherrer equation gives a rough estimate of the broadening caused by crystallite size. For crystal structure analyses the crystallite size should be significantly smaller than about 1 µm in order to avoid line broadening.

For example when Will et al. (1987b) measured fine grained Co_3O_4 they observed considerable line broadening of about twice that of the standard silicon line breadth. They analyzed the individual peaks and determined an average particle size $t(hkl)$ from the observed profile width $B_O(2\Theta)$ in comparison to the silicon profile width of the instrument $B_{Si}(2\Theta)$ using the Scherrer equation

$$t(hkl) = 0.9\lambda / ((B_O^2(2\Theta) - B_{Si}^2(2\Theta))^{1/2}\cos\Theta) \tag{1.11}$$

The size calculated was about the same for all reflections and averaged 235(47) Å indicating the particles had the same dimension in all directions.

A microstructure analysis of line-broadening using synchrotron X-ray diffraction data has been discussed by Huang et al. (1987). Using high resolution synchrotron radiation makes it easy to determine crystalline sizes, microstrains and stacking fault probabilities in materials. The material studied was polycrystalline palladium. Warren-Averbach analysis (Warren and Averbach 1950, 1952) was done with respect to the three major crystal axis [111], [100] and [110].

The subject is treated in detail in the monograph *"X-ray Diffraction Procedures for Polycrystalline and Amorphous Materials"* by Klug and Alexander. Since it requires detailed analysis of the measured peak profiles, if possible of all peaks independently, it is a subject for the Two Stage method rather than the Rietveld or Whole Powder Pattern Modeling.

For a detailed analysis the peak profiles must be analyzed with a Voigt function. Ahtee and his co-workers (Ahtee et al. 1984) introduced and incorporated the actual Voigt function to describe reflection profiles. This function has sufficient flexibility to separate the important parameters. The contribution of crystallite size and faulting is Lorentzian, which reaches out far enough from the peak. They are approximated by a Lorentzian of a width proportional to $1/\cos\Theta$ (Eq. 1.12a). Strain broadening is covered by the Gaussian, which has a width proportional to $\tan\Theta$ (Eq. 1.12b). Instrumental effects arise from geometrical and physical aberrations. According to Wilson (1963) the primary effects on the width of a reflection are proportional to $\cot\Theta$. In summary the full width at half maximum, *FWHM*, of the component functions of the Voigtian can be written as:

$$(2w)_{\text{Lorentz}} = X\tan\Theta + Y / \cos\Theta \tag{1.12a}$$

$$(2w)_{\text{Gauss}} = I_{\text{obs}}(T\cot^2\Theta + U\tan^2\Theta + V\tan\Theta + W)^{1/2} \tag{1.12b}$$

The real Voigt function has been used by Suortti et al. (1979) for the analysis of powder diffraction profiles of Ni measured with X-rays and neutrons. In the outcome they could show that in the neutron case the instrumental factors are dominant, while the X-ray profiles are determined by the size and strain of the particles. As one conclusion it turned out that a profile analysis with a Voigt function can be successfully used only for a profile which can be completely resolved, that means for simple structures with only a few and well resolved reflections. To quote the authors: "in other words the Voigtian refinement can be used only in cases where it is not needed".

Both investigations by Ahtee (Ahtee et al. 1984) and by Suortti (Suortti et al. 1979) are based on a separate fit of individual reflections. But it must be said that the Rietveld method has in principle the potential to determine microstructural disorder parameters in polycrystalline materials. Lutterotti and Scardi (1990) have proposed a procedure for simultaneous refinement of structural and disorder parameters. They included crystallite size and shape and r.m.s. microstrain as fitting parameters, replacing the well known formula of Cagliotti et al. (1958) for the angular dependence of the peak width. Their method has been tested successfully for ZrO_2, $Zr_{0.82}Ce_{0.18}O_2$ and α-Al_2O_3.

Another, quite novel and unusual approach has been proposed and tested by Toraya (1988). He uses a deconvolution procedure applicable to overlapping reflections. The procedure is based on direct fitting of the calculated intensities to the observed ones. The method of direct deconvolution has been developed by Paterson (1950). In contrast to other methods it requires no analytical expression. In this method the integral equation is replaced with a set of linear equations which are solved with the help of the relaxation method. This method in its initial use is stable and applicable only to the deconvolution of isolated peaks. Toraya has developed this new method for deconvolution of overlapping reflections by the direct fitting of the observed and simulated profiles. Then the Gauss-Newton method is adopted to solve the integral equation. The procedure has been applied to the deconvolution of overlapping reflections from ytrria-stabilized ZrO_2. The analysis gave the average crystallite size and microstrain successfully.

Honkimäki and Suortti (1992) have presented a procedure to derive crystallite size and strain by total pattern fitting of energy dispersive powder diffraction data. Data collect by energy dispersive methods, with synchrotron or laboratory X-rays, are usually analyzed in a more elaborate way. Honkimäki and Suortti tested an approach to analyze such data. Difficulties arise for one part, because of unavoidable impurities in the pattern, for the other part because of the background. Impurity peaks of the measured spectra were in this procedure removed before fitting the model. The background coming from incoherent and coherent scattering have been treated by calculating the incoherent part from theoretical cross sections. The coherent part is described as a sum of discrete Bragg peaks and acoustic and optic-phonon thermal diffuse scattering is calculated from Debye and Einstein models, respectively. The method was applied to patterns of Mg, Al and Ti, all compounds with very simple structures and with data available for calculating coherent and incoherent contributions. Summarizing, the analysis gave very good results, however the procedure for fitting is tedious and it must be doubted whether it is widely applicable.

Chapter 2

The Rietveld Method

2.1
The Rietveld Method

2.1.1
The Early Days of the Rietveld Method

Hugo Rietveld came from Australia to Petten in 1964. In Australia, at the HIFAR in Lucas Heights, the overall emphasis was on single crystal diffraction, because at that time powder methods were considered to be inferior to single crystal work. In Petten on the other hand a new group had just been formed and engaged themselves in the construction of a high resolution powder diffractometer. Consequently the emphasis here was on powder diffraction. On the scientific side the main interest in Petten were uranium compounds. The first structures to be studied were rather simple and of high symmetry, with the result that the peaks were more or less well resolved and the powder diffraction data were handled on the basis of single crystal data. Integrated intensities could be obtained and conventional least squares refinement programs could be applied.

When later the compounds became more complex and of lower symmetry the overlapping peaks became a severe problem. As a first step the scientist in Petten improved the resolution of the neutron diffractometer by using a wavelength of 2.6 Å larger than the usually used wavelength of 1.55 Å (and by eliminating higher order wavelengths). For structure refinement the increase in resolution resulted in better defined patterns, but not to such an extent that the peaks were completely resolved. The solution Rietveld was thinking of, was to refine the structure by using not only individual Bragg reflections, but overlapping reflections as a whole. Only groups of reflections were used in his first attempts, not the whole pattern. It worked well, but still extra information contained in the profile was lost. The following step consequently was to separate the overlapping peaks by fitting profile functions. In neutron diffraction the peaks are highly Gaussian, and Rietveld used a Gaussian profile to separate and fit overlapping peaks (Rietveld 1967). It worked well for his first, basically simple structures, but did not work naturally for severe overlaps.

The next step consequently was to consider not just a group of overlapping peaks, but the whole pattern. Rietveld thereby followed a completely new and from earlier methods different approach: He said, why not take the complete diffraction pattern as the experimental data set, take each step scan value $y(i)$ as an observation in a least squares calculation and compare them in a least squares procedure together for the whole pattern. The mathematical basis therefore was the peak profile, and he called his method "profile refinement" (Rietveld 1967, 1969).

Limitations were set by the computers available at that time. Despite the fact that computers had entered crystallography in many respects, they were still to small to handle problems of that size with up to several 1 000 observations. To consider using individual intensities $y(i)$ constituting a step-scanned diffraction peak as data seemed at first completely unrealistic. However with the computers becoming bigger and faster almost every year Rietveld saw a real chance to promote his idea and to use the individual intensities of the whole pattern, at least sections of it. The expression for refinement in the least squares program was $y(i) = \Sigma w(i,k) S(k)^2$ where S is the crystallographic structure factor and w a weighting factor. Background was not yet included in his first refinement program and well resolved single peaks were included as such.

In Petten Rietveld began his career with structure analyses from powder diffraction data, collected with X-rays as well as with neutrons. The main compounds were MUO_4 with M = Ca, Sr and Ba (Rietveld 1966). The refinement was in essence conventional The crystals were measured first with a conventional Philips X-ray powder diffractometer and with a Guinier camera, then with neutrons in order to determine oxygen positions. The X-ray patterns served to determine unit cells, symmetry and the preliminary structural work. For the Ca-compound 43 well resolved peak intensities of the Guinier data were taken to calculate a 3-dimensional Patterson synthesis yielding positions for uranium and three calcium atoms. Three-dimensional Fourier maps improved the positions. Refinement with those data, and also with the neutron data did not lead to the expected results because the data suffered from severe overlaps. In order to overcome this drawback the publication mentions that a least squares program was specifically adapted to these special conditions to obtain intensities from overlapping reflections. There is no mention of his general ideas of full pattern refinement, but we may consider this approach as a forerunner of the Rietveld method. For selected groups of peaks the function to be minimized was defined as $\Sigma(i) \, w(\Sigma(r) jF^2(\mathrm{obs}) - \Sigma(r)jF^2(\mathrm{calc}))^2$ with j the multiplicity of the reflections, w its weight, $\Sigma(r)$ the sum of the overlapping reflections which were treated as one intensity. $\Sigma(i)$ is the sum of all measured intensities. The special program used single well resolved peaks as well as those overlapping reflections since the computer, the Electrologica X1, then available in Petten, was not powerful enough to solve a least squares problem of more than a very limited number of data and of parameters. This paper was submitted to *Acta Crystallographica* in May 1965, one year before the IUCr congress in Moscow.

A first mention of a full pattern refinement was in 1966 at the IUCr congress in Moscow where he gave a first report on his method. "There was hardly any response" (his own words). Moscow was focused on the fast growing field of single crystal structure analyses, especially, since G. Hauptmann had just received the Nobel Prize (together with Karle and Karle) for his direct method approach: "It is interesting, that a mathematician receives the Nobel Prize in chemistry for his work in crystallography" (Hauptmann's own words in Moscow). Finally in 1967, one year after Moscow, a short note appeared: *"Line profiles of neutron powder-diffraction peaks for structure refinement"* (Rietveld 1967). It describes the much simpler problem of a "profile refinement" technique, a forerunner to the actual full pattern Rietveld refinement following two years later. To test the method the structure of WO_3 was refined, which had been studied previously with X-rays (Loopstra and Boldrini 1966). This short note describes what we call today the Rietveld refinement method.

A full paper with the title "*A Profile Refinement Method for Nuclear and Magnetic Structures*" was published in 1969. The inherent presence of overlapping reflections generally prevents the full use of the available information of a powder diffraction pattern to refine the structural parameters. For the full use the profiles of the peaks have to be included. At that time computers were the limitations in Rietveld's approach. In order to limit the number of data, e.g. of observations in the least squares program, several single peak intensities were replaced by a δ-peak with all $y(i)$ in this area replaced by zeros except for one value made equal to the area of the Gaussian peak. It worked very well. The observed diffraction pattern and its calculated profiles are shown and they are very convincing. The achieved R-value is not given. This first version of the program was written in machine language for the computer Electrologica X1. It made the program difficult to use outside Petten. But it was his idea, which deserves full recognition and which meant an enormous step forward. Also this distinguishes his approach fundamentally from Will's "Two Stage" approach a few years earlier.

The first full paper on the Rietveld method appeared by Rietveld in 1969 (Rietveld 1969). In this paper a comprehensive outline of his method is given discussing almost all of the inherent problems of full pattern analysis: peak shape, peak width, asymmetry, preferred orientation, the parameters and their refinement. At that time the program was already written in ALGOL60, a computer language favored in Europe at that time and more convenient than the original Electrologica machine language. This structure refinement method does not use integrated (neutron) intensities, single or overlapping, but employs directly the profile intensities obtained from step-scanning measurement of the powder diagram. A number of structures had already been refined with his new method and the paper lists 12 structures, all from neutron diffraction data, which had been successfully refined with his program. The authors were all scientist from Petten. Detailed discussion is given for two examples, the nuclear structures of $CaUO_4$ and of Sr_2UO_5. $CaUO_4$ had been studied preciously (Loopstra and Rietveld 1969) and served as a test for the full pattern refinement. The diffraction pattern of $CaUO_4$ shows hardly any overlap, and the structure had been refined successfully with integrated intensities. The fit between observed and calculated patterns shows an almost perfect fit and it demonstrates the power of this method. The profile, e.g. the peak shape was taken as Gaussian. The second example of Sr_2UO_5 is a pattern with severe overlaps. At higher angles up to ten reflections may contribute to one profile intensity. Again the agreement between observed and calculated intensities $y(i)$ is convincingly good. These two examples demonstrated the potential of the method for refining crystal structures from powder diffraction patterns.

The conclusion to be drawn from these two examples is, that the inherent drawbacks of the powder method, i.e. the loss of information as a result of overlap, can be effectively overcome by Rietveld's method, and powder diffraction can obviously compete with single crystal methods, at least in principle. These first applications were based on neutron diffraction data with it convenient and easy to program Gaussian peak shape. But Rietveld realized already at that stage, that the method could just as well be applied with X-ray diffraction data if a satisfactory function could be found to describe the peak profiles. "Due to the lack of an actual problem, I did not pursue it any further" (his own words).

We may say, that the breakthrough of this method came in 1978 at the congress of the IUCr in Warsaw. Following the main congress a satellite meeting was organized in Krakow, devoted especially to powder diffraction. It was then suggested by Ray Young that this method, generally known as "full pattern refinement" should in future be called "Rietveld refinement". A full acceptance followed at an international workshop on the Rietveld method held in 1989 in Petten.

Already in 1989 the program was available in the general accepted language FORTRAN IV. This was the basis for the final acceptance of his program and method. When Cheetham and Taylor (1977) published a comprehensive description of the Rietveld method in 1977 referring to neutron diffraction data, basically meant for chemist, they give a list of about 172 structures refined this way with neutron diffraction data. At that time the program was mainly used to refine structures from data obtained by fixed wave length neutron diffraction. After 1977 the method was extended by a number of scientist to X-ray data and was finally generally accepted. Today a considerable number of versions of the "Rietveld program" are in use with quite different aspects including lately even the use in quantitative analysis.

A word of caution:

i   to the nomenclature: The Rietveld method is a least squares refinement procedure where the experimental step-scanned values are adapted to calculated ones. The profiles are considered to be known, and a model for a crystal structure available. With years and the steadily improving of the programs the number of parameters to be refined has been increased dramatically. In some versions profile parameters are included. However we must keep in mind, the Rietveld procedure still is a refinement of structural, and sometimes instrumental and profile parameters. This is different from the Two Stage method, where the profile, and the instrumental parameters are determined and refined at the beginning of a study, and then kept constant during structure refinement.

ii   to the temperature factor: The reliability with which temperature factors can be determined is debatable. Temperature factors are especially sensitive to high angle data, and for this reason neutron diffraction is in general more reliable than X-rays, where the patterns are hampered by the fast fall-off due to the scattering form factors with angle, and more over are strongly influenced by bonding features. The inclusion of anisotropic temperature factors, as it is provided in most Rietveld program versions, adds to the list of parameters and consequently reduces the R-values, but give hardly a description of the thermal movement of the atoms.

iii to the R-indices: The profile R-index rarely falls to the value expected on the basis of counting statistics. The primary causes of this are errors in the background estimation and the deficiencies of the assumed peak shapes. As a result it is often difficult to know when the refinement procedure is really complete. It is recommended to calculate difference maps and to take deviations serious. Also a difference map serves the purpose of a reliable refinement better than numerical R-values, which are discussed in many papers on several formulas.

2.1.2
The Method

The Rietveld method is a complex minimization procedure. It is not an active tool for *ab initio* crystal structure analysis. It can only slightly modify a preconceived model built on external previous knowledge. The starting parameters for such a model must be reasonable close to the final values. More over, the sequence into which the different parameters are being refined needs to be carefully studied. This is discussed in detail in Sect. 2.1.5. The Rietveld method is a structure refinement procedure. It uses step intensity data $y(i)$, whereby each data point is treated as an observation.

The idea behind the Rietveld method is to consider the entire powder diffraction pattern using a variety of refinable parameters. That way the intrinsic problem of any powder diffraction pattern with its systematic and accidental peak overlaps is overcome. It is the intention to extract as much information as possible from a powder pattern. At the beginning this was restricted to atomic positions from neutron diffraction patterns (Rietveld 1967). The first full publication has the title *"A Profile Refinement Method for Nuclear and Magnetic Structures"* (Rietveld 1969). There he describes a structure refinement method (and program) which is not based on integrated intensities, as is done in single crystal structure refinements. His program employs directly the individual intensities $y(i)$ at each 2Θ value obtained from step-scanning measurements of powder diffraction patterns. This was a novel and unusual approach. Rietveld considered such an approach and such a program since investigations based on powder methods had gained new importance, especially in neutron diffraction, owing to the general lack of large specimens for single crystal methods.

Diffraction of a polycrystalline sample, with neutrons or X-rays, reduces the three-dimensional reciprocal lattice to a one-dimensional diagram. As a consequence such patterns suffer from overlapping peaks, sometimes accidental due to a lack of resolution, sometimes intrinsic in patterns for samples with cubic or trigonal symmetry. Especially at higher diffraction angles, in low symmetry structures or when large unit cells are involved this is a serious problem. It is inevitable that certain information is lost. By using the step-scanned profile intensities instead of integrated intensities in the refinement procedure this difficulty can be overcome to a great extent allowing at the same time the extraction of a maximum amount of information. Rietveld's approach is based on the complete diffraction pattern, including background. The method is called originally "full pattern refinement". Since the IUCr satellite meeting on powder diffraction in Krakow, Poland, in 1978 terms like "Rietveld refinement", "Rietveld method" or "Rietveld analysis" were suggested. Rietveld method is generally and widely used today and extended to more and more fields of analysis based on diffraction, like phase analysis or texture analysis. This approach is fundamentally different from the Two Stage method based on POWLS, which was written with the same scientific background in the early sixties before the Rietveld program became known. POWLS requires intensities, as peak maxima or as integrated intensities. Later in the eighties the Two Stage method was extended by adding as a first step profile fitting resulting in a complete separation of the diffraction peaks. This step yields reliable integrated intensities comparable to single crystal data, and this makes this method more general and closer to single crystal analyses. This method consequently is called today the "Two Stage method".

There is much more information hidden in a powder pattern than just atomic positions, site occupancies and Debye-Waller factors. To name a few: lattice parameters and space group can be deduced and refined from the peak positions of the reflections; the amorphous fraction in the specimen or local order/disorder can be deduced from the background; particle size, strain/stress and domain size in the sample from analyzing the broadening of the peaks, *FWHM*, and in the recent developments qualitative and quantitative phase analysis. Special version have been written and numerous versions of the program are available today.

To understand the Rietveld method, or any method based on powder diffraction data several inherent properties and problems have to be discussed:

i   Peak shape
ii Peak width (*FWHM*, full width at half maximum)
iii Preferred orientation
iv Method of calculation

Peak Shape

The measured profile of a single, well resolved powder diffraction peak is dependent on two intrinsic parameters: (*i*) An instrumental parameter including the spectral distribution, i.e. the monochromatic mosaic distribution, and the transmission function determined by the (Soller) slits, and (*ii*) the sample contribution based on the crystal structure and the crystallinity of the sample. While these contributions can have a form not necessarily Gaussian, it is an empirical fact that their convolution produces in neutron diffraction patterns almost exactly a Gaussian peak shape. This is different in X-ray diffraction, where especially the instrumental contributions lead to rather complicated peak profiles. A number of profiles have been suggested and tested in the past and some are still preferred by most scientists. The majority however just uses the so-called Pseudo-Voigt function, the summation of a Lorentzian and Gaussian function with adjustable contributions, e.g. parameter η. (The real Voigt function is the convolution of these two functions). Often this Pseudo-Voigt function is split in a left hand side and a right hand side of a diffraction peak to accommodate asymmetries in the diffraction peaks.

The Peak Width

The width of the diffraction peaks is the second important parameter and variable when describing a diffraction pattern. The peak width, described as "full width at half maximum", *FWHM*, is in general a function of the diffraction angle 2Θ. Cagliotti has studied the angle dependence of *FWHM* for neutron diffraction (Cagliotti et al. 1958). They have given a formula to describe this angular dependence.

$$(FWHM)_k = U\tan^2\theta_k + V\tan\theta_k + W \tag{2.1}$$

with U, V and W adjustable parameters. This simple formula describes adequately the experimentally observed variation of half width with scattering angle. Today it is also used for X-ray and synchrotron diffraction patterns. The parameters are refinable quantities in Rietveld's least squares calculations. The initial and approximate starting values are found at the beginning of an experimental cycle by measuring *FWHM* in a standard sample with individual single peaks. These values are unchanged as long as the experimental setup is unchanged and if there is no line broadening coming from the crystallites.

Preferred Orientation Correction

Preferred orientation is always a problem when dealing with polycrystalline samples. It was already recognized by Rietveld and incorporated in his first programs. It is discussed in detail in Sects. 1.2.9 and 3.4.6.

Method of Calculation

For Rietveld refinements the data must be in digital form. This is common today with the usual quantum detectors, with energy dispersive detectors and even with (digitizing) micro-densitometers. The basis for the refinement are the numerical intensity values y_i at each of several thousand equal steps along the scattering angle 2Θ, with increments $\Delta(2\Theta)$. Typical step sizes range from 0.01 to 0.05°, in some cases (with synchrotron radiation) even to 0.001° with the consequence of course of long counting times. The best fit sought is the best least squares fit to all of the thousands of y_i simultaneously. The quantity minimized therefore is in general terms

$$S = \sum_i w_i(y_i(obs) - y_i(calc))^2 = \text{Minimum} \tag{2.2}$$

with w_i the weight of each observation point, $y_i(obs)$ and $y_i(calc)$ the observed and from a model calculated intensities at each step. The sum i is over all data points. Background is assumed to be subtracted.

Using the Rietveld programs needs experience despite its simple approach and its advertised ease in using it. The least squares calculation contains a considerable number of parameters, actual too many to refine at one time. They are often affected by considerable correlations. A bloc refinement, e.g. a separate refinement of groups of parameters is advised.

The main parameters in today's programs can be divided into three groups: The first group defines basic experimental parameters: the profile parameters, the halfwidths, possible asymmetries of the diffraction peaks and a zero point adjustment:

U, V, W Halfwidths parameters (Eq. 2.1)
Z Counter zeropoint
P Asymmetry parameter

The second group contains the unit cell parameters, and the crystallographic symmetry, especially space group. The unit cell parameters are with its general definition:

A, B, C, D, E, F with $1/d^2 = Ah^2 + Bk^2 + Cl^2 + Dkl + Ehl + Fhl$
G Preferred orientation parameter

Finally the third group contains the actual structural parameters:

c	Overall scale factor with $y(\text{calc}) = cy(\text{obs})$
Q	Overall isotropic Debye-Waller parameter (this value is not needed and seldom used; it is consumed in the individual atomic Debye-Waller factors
x_j, y_j, z_j	Fractional positional coordinates of the jth atom in the asymmetric unit
B_j	Atomic (isotropic) Debye-Waller factors (anisotropic parameters can, and have been included)
n_j	Occupation number of each crystallographic site

with j the jth atom in the asymmetric unit. In the least squares refinement the problem is not linear in the parameters, in Rietveld programs as well as in POWLS, therefore approximate values for all parameters are required for the first refinement cycle. In subsequent refinement cycles these parameters are refined until a certain convergence criterion is reached, or the refinement is stopped by the operator.

Background

The background is a crucial point in the refinement. Equation 2.2 assumes that the background has been subtracted. Background becomes more and more important since more and more specimens have high background or contain amorphous material, whose scattering shows up in the background. In practice it is often difficult to define the background in the diagram. The background is mostly defined by footing marks between the diffraction peaks, where no contribution occurs from the specimen. Since the peaks are often not well separated it becomes difficult or ambiguous to set such footing marks. Nevertheless in most programs background is included as one or several extra parameters. But the question remains, what about amorphous contributions.

2.1.3
Rietveld Refinement with X-Ray Powder Diffraction Data

Rietveld's own work was on neutron diffraction data. He did foresee the application also for X-ray patterns, but he never applied it "due to the lack of an actual problem". The method found its way into X-ray diffraction much later and quite slowly. Today this has changed. The method finds its main use and success in X-ray work, especially with data collected at synchrotrons. Young et al. summarize it in 1977 (Young et al. 1977). The reason why scientists hesitated for a long time to apply the method to X-ray patterns were the considerable more complex profiles in X-ray patterns, especially when collected with sealed X-ray tubes in the laboratory. These profiles are quite different from the simple Gaussian profiles in neutron diffraction. Also in X-ray patterns the profiles vary strongly with scattering angle in a way quite different and much more complicated than in neutron diffraction patterns, where the variation of halfwidth with angle can be described by a simple formula as determined by Cagliotti (Cagliotti et al. 1958).

The difficulties in X-ray patterns are several: The scattering of X-rays is by the electrons and is governed by angle dependant atomic scattering factors, including anomalous dispersion. The data are further affected by the Lorentz and polarization factors, which become even more complicated if monochromators are used in the primary beam. Often absorption corrections are necessary, and the alpha doublet has to be handled. These factors are basically non-existent in neutron diffraction. A great step forward came, when synchrotron radiation became available on a broader basis with its strictly single monochromatic radiation and a nearly 100% polarization of the beam in the accelerator plane.

The main obstacle are the profiles of the peaks. A number of profiles have been used by scientists and incorporated in the programs. In Sect. 1.2.5, and Table 1.5, the different profile functions are discussed. Today the Pseudo-Voigt function, a partial addition of a Gaussian and a Lorentzian function, is the most common function used. As mentioned the profile is highly dependant on the scattering angle with an angle dependent splitting of the $\alpha(1)/\alpha(2)$ contributions when using sealed tubes in laboratory experiments. This leads to a significant splitting and even a total separation of the two components, which can be clearly seen at higher angles. Successful works to overcome this problem and to handle the splitting are found for the first time in papers by Huang and Parrish (Huang and Parrish 1975) and by Parrish and his co-workers (Parrish et al. 1976).

Background in X-ray diffraction patterns is another serious problem. There are angle dependant contributions to the background coming from fluorescence and air scattering and leading to a non-uniform background. Amorphous contents in the sample give further non-uniform contributions to the background. Therefore background must be dealt with carefully before a successful refinement. The general procedure today is to set footing marks between the peaks and use those to calculate or subtract the background. This may become difficult and very often it cannot be determined straight forward, especially if the peaks are too close to each other with considerable overlap, for example if larger unit cells or lower symmetries are dominating. This leaves little unambiguous background between the peaks.

Wiles and Young (1981) have written *ab initio* a new Rietveld refinement program, DBW3.2, which is generally used today. Since Rietveld refinements are used today predominantly with X-ray or synchrotron data, this program shall be described briefly: The program has been written in FORTRAN IV and contains about 5 000 FORTRAN statements. For convenience it has, just like POWLS, built-in a direct applicability with all space groups and scattering factors for all elements, either numerical or as coefficients of an exponential description, plus anomalous scattering corrections and also nuclear scattering lengths when used with date from thermal neutron scattering. The space group is entered with the standard space group symbol from the *International Tables of Crystallography*. The program then generates the multiplicities.

The quantity minimized is

$$S = \sum_i w_i (y_i(obs) - y_i(calc))^2 = \text{Minimum} \tag{2.3}$$

with w the weight given by

$$w_i^{-1} = \sigma_i^2 = \sigma_{ig}^2 + \sigma_{ib}^2 \tag{2.4}$$

with σ_{ig} the standard deviation, normally based on counting statistics, and σ_{ib} for the background at the ith step. The y_i(calc) values are the intensity contributions calculated by the usual crystallographic formulas for structure factors $F(hkl)$ and intensities including Lorentz and polarization factors and also a factor P or G to correct for preferred orientation. Background may be included or subtracted before the refinement. Several profile functions are contained in the program.

The parameters that can be adjusted in the least squares refinement, in principle simultaneously, include:

- Lattice parameters $(a, b, c, \alpha, \beta, \gamma)$
- Atomic positions (x, y, z)
- Atomic site occupancies
- Atomic thermal vibrational parameters, isotropic or anisotropic
- Profile including U, V, W from the Cagliotti formula and asymmetry
- Preferred orientation
- Background function
- 2Θ-zero correction
- Overall scale factor
- Overall isotropic thermal B

As input information the program requires:

- Initial values of all variable parameters (listed above)
- Step-scan data in equal increments 2Θ
- 2Θ limits, starting and ending values of 2Θ, and regions which shall be excluded in the data
- Wavelength data

Some examples are given in Sect. 1.2.6. The structures listed were refined with Rietveld programs after refineable models had been determined by other conventional crystallographic methods. One specific example, synthetic hercynite, $FeAl_2O_4$, a mineral, shall be mentioned here, because the structure was solved directly from the X-ray powder diffraction data with the Rietveld procedure without pattern decomposition (Hill 1984). Hercynite crystallizes with the spinel structure, which could be used as input for a model structure. The structure has only three positional parameters. A flexible profile shape of the Pseudo-Voigt type was used with 33.7 to 99.8% Lorentzian character as 2Θ increased. The structure refinements were undertaken with the full-profile Rietveld-type program DBW3.2 (Rietveld 1969; Wiles and Young 1981). Such examples of direct Rietveld refinement are especially useful in solid state chemistry, where a series of compounds is synthesized with structures known from similar compounds, and where starting parameters can be safely guessed from experience.

In a comparison of different profile shape functions by Young and Wiles (1982) it could be shown, that positional parameters are generally not affected significantly by the choice of the function used. Thermal parameters however often were and they are all together not reliable quantities for description of thermal motions of atoms in the unit cell. The overall best criteria for the fit are difference plots between observed and calculated data.

Criteria for a successful refinement are the following:

- A difference plot $y_i(\text{obs}) - y_i(\text{calc})$
- No maximum deviation at any point in the difference plot
- A low R-index R_{wp} for the profile fit
- The R-index R_B (B for Bragg) for the agreement of integrated peak values (The values $I(\text{obs})$ are calculated on the basis of the calculated ones and are therefore not directly comparable to single crystal R-indices)
- Structural parameters and their standard deviations (if possible in comparison to similar single crystal results)

R-indices in general are of limited values in judging the results obtained. Powder diffraction data are not directly comparable to single crystal data, but the R-indices used are close to the single crystal structure R-value. Also the standard deviations σ cannot be taken at face value.

Of the several R-indices used to check on the quality of the refinement R_{wp} is statistically the most meaningful indicator since the numerator is the residual that is minimized in the least squares procedure. Any R-index presented is meaningful only if the background has been subtracted. This deserves a word of caution. The influence of background on R-values has been discussed by Jansen E. et al. (1994a) (see also Sect. 3.7.1). The contribution of a high background enters the denumerator and thus will greatly affect the R-indices, e.g. lower the numbers.

R_B, the so-called Bragg R-index, implies a comparison of integrated intensities similar to single crystal refinements. In a Rietveld refinement there are no real integrated intensities and R_B is based on somewhat fictitious observed intensities. It is calculated by allocating the actual observed (step scanned) intensities $y_i(\text{obs})$ to Bragg intensities on the basis of the calculated intensities. Nevertheless it is a useful indicator.

All R-indices are greatly affected, if the crystallites are not ideally "imperfect", e.g. if they are affected by crystallite size and/or microstrain problems. Cryotallite-size shows up in line broadening and produces intrinsic contributions to the Lorentz profile. Microstrain on the other side produce Gaussian shaped profiles. Such contributions will show up in general in the difference plots. Very general the R-indices as quantities for fit and accuracy must be taken with care. One might assign in such cases some significance to η in the Pseudo-Voigt formula, particularly to its angular dependence. This is a good reason to include η as a parameter in the refinement.

2.1.4
Critical Assessment of the Rietveld Method

Structure analysis and structure refinement with powder diffraction data is a true alternative to single crystal measurements, but it is also a challenge. How reliable and how good are the structural data derived from powder data in general and also in comparison with single crystal studies. How good are results gained from full pattern refinements in comparison to profile decomposition analysis. These questions concern primarily positional parameters and their standard deviations, and to a lesser degree the Debye-Waller factors.

The Profile Refinement Method (Rietveld Method)

The profile refinement method, or full pattern analysis, is extensively used today. If the peaks in a diffraction pattern are resolved it is the logical way to determine the structural parameters by conventional refinement directly from the integrated intensities of the separated peaks derived by pattern decomposition or other integration means. This way can, and has been used also if a small number of overlapping peaks occur, then for example also by considering a group of overlapping peaks as a single observation. If the number of overlapping peaks is large the advantage of profile refinement, e.g. Rietveld refinement is clear.

In this case each point on the profile is considered as a single observation y_i which may contain contributions from different Bragg peaks. And here come the problems. The power of a profile total pattern refinement in the analysis of a fairly complex pattern has been, and still is judged primarily by the values of standard deviations assigned to the structural parameters. The quantity to be minimized in the least squares refinement is

$$M = \sum_i w_i (y_i - K \sum_k I_k G_{i,k})^2 \tag{2.5}$$

Here each point is regarded as a separate observation with its own statistical uncertainty and its own weight in the least squares refinement. w_i is the weight of each separate step observation, which should be based on the background corrected intensity y_i. Σk is the sum over different Bragg peaks, K is the scale factor and G represents the profile function centered at $2\Theta_k$.

The Rietveld refinement method has been used in the beginning mainly to the analysis of neutron diffraction powder patterns. Here the profiles are basically Gaussian and no serious problems are encountered. For X-ray data with the much more complicated profile description the question was asked correctly do the profile functions have influence on the resulting parameters. The formula (Eq. 2.5) assumes that the intrinsic diffraction profile G of the crystallites is independent of structural parameters. This is valid for most materials. It may however not necessarily be valid if intrinsic line broadening is significant, for example with small particles or with internal strain in the particles.

Validity of Estimated Standard Deviations

There is a fundamental difference between the two methods. In the Rietveld refinement a least squares fitting is applied to a great number of individual measurements whereas in integrated intensity refinements it is applied only to the integrated measurements of separated peaks, a limited number. This results in a large difference of "observations" when e.s.d. and R-indices are calculated. Here the "number of observations" enter the calculation. It is a strong believe that the estimated standard deviations σ of the refined parameters provided by Rietveld programs are invalid.

In the Rietveld refinement each observation is assigned a statistical weight

$$w_i = 1 / (\sigma^2(y_i) + \sigma^2(B_i)) \tag{2.6}$$

where σ^2 is the variance of the appropriate quantity. Since B_i (B for background) is obtained by graphical means its variance is not known and it is often set arbitrarily to zero. This then gives

$$w_i = 1 \,/\, y_i \tag{2.7}$$

The omission of the variance of the background results in an overestimation of the weight, which does not occur in a corresponding single crystal, or single peak analysis, where the background is usually well determined. To consider this deficiency researchers sometimes include a constant value C, for example $C = 100$, into the weight of each intensity:

$$w_i = 1 \,/\, (y_i + C) \tag{2.8}$$

The controversy of the correctness of the standard deviation was addressed by Sakata and Cooper (1979) and later again by Cooper in 1983 (Cooper 1983). The Rietveld method follows the practice of most crystallographic least squares analysis methods in minimizing the sum of squares of the residuals derived from the observed intensity for each individual measurement. In Rietveld refinements the observed quantities are then the step-scan intensities y_i with innumerable more "observations" than there are actually available.

The standard deviations σ_i of a parameter x_i is given in the least squares refinement by

$$\sigma_i^2 = A_{ii}^{-1} M \,/\, (N - P) \tag{2.9}$$

where A_{ii}^{-1} is the diagonal element of the inverse normal matrix corresponding to the ith parameter, M is the quantity minimized, N the number of statistically independent observations and P the number of independently varied parameters.

Powder refinements are usually, almost necessarily, not as good as single crystal refinements because individual reflections are not observed. The definition of the number of data points being the number of observations is artificial in the Rietveld method, because they can be varied at will from zero to infinity by varying the step size. The e.s.d. become zero at infinity. In single crystal data the observations are the intensities of each reflection (hkl). The estimated standard deviation of the refined parameters in these cases are calculated using the standard expression involving the diagonal elements of the inverse normal matrix.

Comparison with Pattern Decomposition Data

The differences between Rietveld refinement and integrated intensity refinement methods are not well understood. Since the peak shape can be assumed to be well defined it is reasonable, it is necessary, that the two methods give the same structural values. However because of the weighting scheme adopted in the profile analysis method the values of the structural parameters will not necessarily be the same as those obtained from integrated or single crystal intensity measurements. It is therefore not obvious that these quantities can be compared directly with corresponding quantities from conventional single crystal data. Comparative studies are necessary and several can be found in the literature.

Taylor (1987) has presented already in 1987 a comparison of Rietveld refinement vs. profile decomposition in some detail. In one compound X-ray powder data for H^+-ZSM5, a zeolite, are compared to single crystal diffraction data. From the powder diffraction pattern individual intensities and structure factors for each reflection were extracted by a point-by-point decomposition of the profiles for each (hkl) reflection by the Rietveld (1969) decomposition formula. Ninety-two integrated intensity values were obtained. These data were refined with a conventional single crystal least squares program. It was noticed that the e.s.d. of the Si and O atoms were several times smaller than those obtained by Rietveld or single crystal analysis.

A second comparison (Taylor 1987) concerned three compounds measured by neutron diffraction: cubic K_2NiF_6, second a moderately overlapped pattern of monoclinic UI_4, and thirdly a very overlapped triclinic $MoOCl_4$. These three compounds were refined both by the two-stage profile decomposition method and by Rietveld refinement. Statistical weights were used for the Rietveld refinement, unit weights for the two-stage refinements.

The comparison shows that the structural parameter errors became less for the decomposition refinement as the overlap in the pattern increases. The two methods are fundamentally different in the case of the structural parameters. The pattern decomposition yields intensity values which resemble single crystal data, and should not be completely artificial if the decomposition process is effective. The results should approach the single crystal data set. The Rietveld data are based on a set of step-scan observations, which is quite removed from a single crystal data set. Also, and much more serious, the number of observations for Rietveld refinement is arbitrary, as the step-width is variable. On the other hand, the number of observations in pattern decomposed refinement is fixed at the (hkl) population. Therefore, if the decomposition yields good integrated intensities these can be properly weighted and there may be consequently an advantage in the decomposition method because of the better data distribution.

Comparison with Single Crystal Data

There are also reports on the comparison of Rietveld refinements with single crystal data. One such comparison has already been mentioned in Sect. 1.2.5, where Hill and Howard compared three compounds measured with neutron diffraction with single crystal data.

A direct and very detailed comparison has been published by Bernard et al. (1991b). They investigated $Cd_5(OH)_8(NO_3)_2 \cdot 2H_2O$, a rather complex compound. The powder diffraction data were collected on a Siemens D5000 powder diffractometer operating with strictly monochromatic copper $K\alpha_1$ radiation. The diffraction pattern was decomposed into its Bragg reflections with the Siemens DIFFRAC-AT software using a Pseudo-Voigt function. The shape factor was $\eta = 0.98$ indicating a strong Lorentzian character of the line shape. From the precise powder data the automatic indexing program DICVOL gave the lattice parameters.

The structure was determined from a 3-dimensional Patterson map using 96 unambiguously indexed reflections from the powder data set. Twenty-three unobserved, but space group allowed reflections with small, but non-zero intensity were included in the calculation of the Patterson map. Those 23 reflections are meaningless in a Patterson (or any Fourier) calculation, since they will contribute zero, e.g. only noise to the map. This is different of course in the least squares refinement, where zero observations are valuable information and must lead to zero values in the calculation. With follow-up least squares and Fourier calculations the complete structure could be determined. The approximate atomic positions were then used for a least squares structure refinement by the Rietveld method using the program DBW3.2S. The refinement used 444 powder reflections. It included 40 parameters, 19 atomic coordinates, 5 isotropic temperature factors, the scale factor, 4 cell parameters, a zero-point parameter, an asymmetry parameter and 6 parameters to define the functional dependence of the background. The refinement resulted in a only partly satisfactory R-index value of $R_B = 0.083$ (B for Bragg), and a profile value $R_p = 0.093$ and $R_{wp} = 0.116$. It should be mentioned, that the powder diffraction data for this study were collected in the laboratory with a standard diffractometer.

With $a = 18.931$ Å, $b = 6.858$ Å, $c = 5.931$ Å and $\beta = 94.85°$, monoclinic symmetry and space group C2/m the structure is rather complex. It results in a heavily crowded powder diffraction pattern. The final difference plot shown in the publication is fairly good, but it must also be realized that the measured data beyond $2\Theta > 60°$ disappear in the background and can hardly be made responsible for a contribution to the refinement.

The single crystal data were collected on a Enraf-Nonius CAD-4 diffractometer. A total of 2 504 independent reflections were collected, from which 2 218 were larger than $3\sigma(I)$. 444 reflections corresponded to the same $\sin\Theta/\lambda$ limit as in the powder case and those were used in the comparative refinement.

Overall we must consider these studies a success. Crystal structure analysis with powder diffraction data is certainly a useful way out if single crystal are not available. This last example is in particular worth mentioning since this study showed the ability to solve crystal structures with more than moderate complexity with powder diffraction technique. The result reveals a remarkably well agreement with the single crystal data; the parameters, unit cell as well as atomic positional parameters between powder and single crystal data are in very good agreement. The precision of the positional parameters is lower by a factor of about 10 on average in the powder study. This has been observed, and discussed, in a number of similar studies and several authors have observed a similar trend in comparative studies (see for example Attfield at al. 1986). It may be asked whether the precision quoted in Rietveld refinements using powder diffraction data are really meaningful. The resulting atomic parameters certainly serve their purpose and are sufficient for normal and most purposes.

In one further study a comparison was made with Fourier maps calculated from single crystal data and with structure factors from powder diffraction derived by pattern decomposition (Will et al. 1988b). This study is not directly an assessment of the Rietveld method, but nevertheless it proves the power of powder diffraction. In this investigation electron density sections have been compared directly with electron density maps calculated from high precision single crystal data. The result is discussed in more detail later in this monograph in Sect. 3.10.1.

Comparison of the Rietveld Method with the Two Stage Method

Both methods are based on powder diffraction data. They use a different approach to determine crystal structure parameters. In the end both ways should give the same results. A detailed comparison has been presented in a paper by Will et al. (1983a). The merits of the Rietveld method are well known and are very successful in straight forward and conventional structure refinements, when a structural model and the unit cell and symmetry is approximately known.

The Two Stage method on the other hand is superior to total pattern methods if we have unconventional diffraction techniques, high pressure research with, for example, large amounts of contaminating peaks, or in the study of magnetic structures. In this publication the two methods are compared (*i*) for a simple compounds, quartz, measured with synchrotron radiation, and (*ii*) for several more complicated neutron diffraction data.

For quartz two data sets were available, with $\lambda = 1.0020$ and 1.2823 Å, both measured with synchrotron radiation. The background was close to zero. Both data sets were analyzed with the Rietveld method and with the Two Stage method using FULFIT for extracting integrated intensities and POWLS for structural refinement. There was no significant difference in the results obtained by the two methods. The results were in both cases in excellent agreement with single crystal data (Levien et al. 1980). The standard deviations from the powder data are about three times larger than the ones from the single crystal data. Rietveld and POWLS resulted in about the same e.s.d. Figure 2.1 depicts selected parts of a quartz pattern for $\lambda = 1.0020$ Å including the profile refinements using FULFIT (above) and the Rietveld routine (below).

A second set of comparison was made with neutron diffraction data, for Li_2CuO_2, UFe_4Al_8, $ThFe_4Al_8$, and $TbNiC_2$. Because of the Gaussian peak shape the Rietveld routine works extremely well and is to be in such simple cases to be preferred to pattern decomposition. Also here some patterns had severe peak overlaps, and in such cases the Rietveld method is much easier – and faster – to use.

The situation is different if magnetic structures, especially incommensurate magnetic structures have to be analyzed In such extreme cases only a Two Stage approach will lead to meaningful results. Examples are discussed in Sect. 3.7.3.

2.1.5
Guidelines for Rietveld Refinement

Structural refinement using the whole pattern or Rietveld method is a powerful technique for extracting structural details from powder diffraction data. With present methods structures up to 200 structural parameters can be refined successfully, if care is taken, and if the data are of sufficient high quality. It is, however not a straight forward simple way, especially for a scientist not familiar with this technique. On the other hand for the experienced scientist it is certainly useful and fairly easy to use. Nevertheless certain pitfalls must be avoided.

Rietveld programs today are sophisticated and highly automated and are claimed to be very "user friendly". This relates the impression that it is easy to use them and in a straight forward procedure to determine and refine a crystal structure from powder

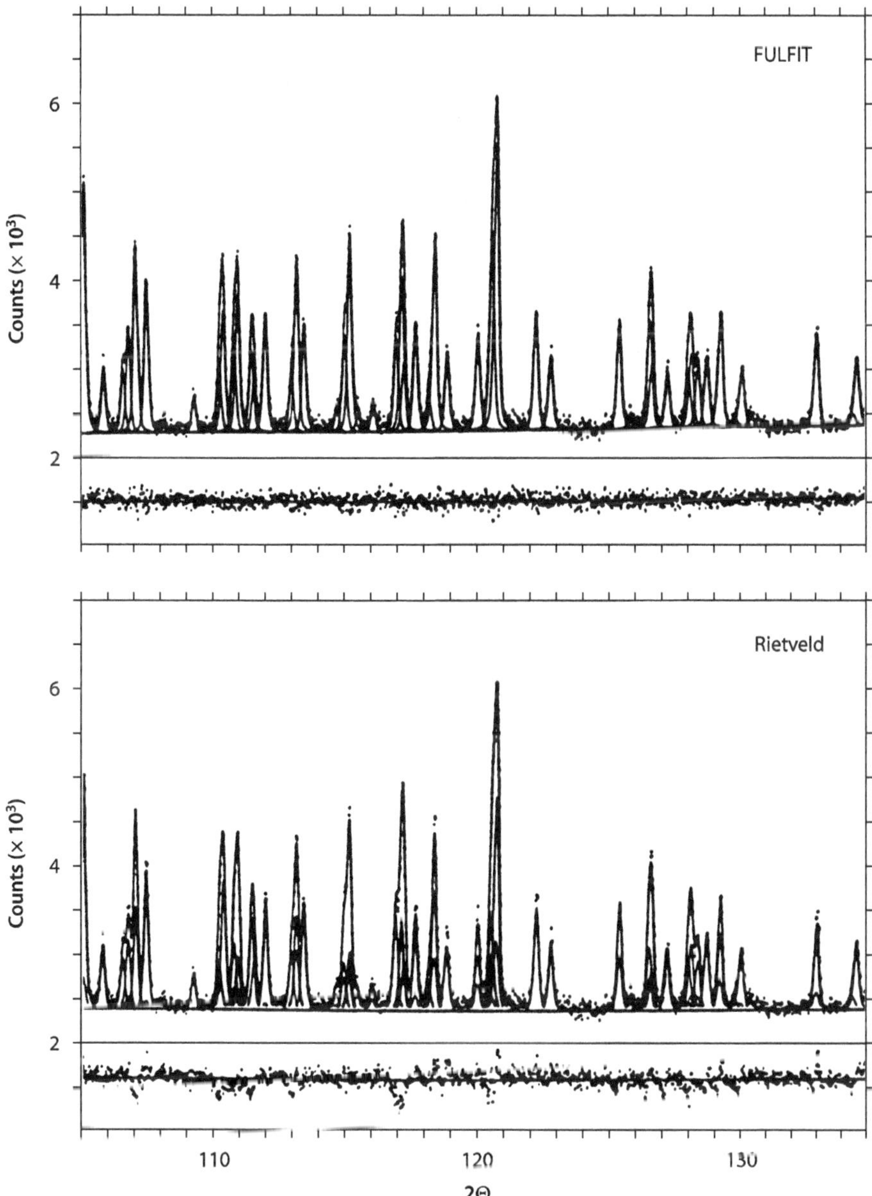

Fig. 2.1. Selected parts of a quartz pattern for $\lambda = 1.0020$ Å including the profile refinements using FULFIT (*top*) and the Rietveld routine (*bottom*)

diffraction data: Take a diffraction diagram, feed the data into a Rietveld program, and that is it. In reality crystal structure analysis based on powder diffraction data needs experience and detailed and deep knowledge in crystallography. The Rietveld programs offer many, too many variables for refinement which, if refined simultaneously, may and often do suffer from strong correlations of the parameters resulting in unacceptable values and erroneous standard deviations. Especially the R-indices given by the programs are very often a source of wrong confidence. They must be looked at carefully. The whole refinement is heavily dependent upon the experience of the user, where crystallographic background and knowledge are certainly needed. This guidelines should help. It can be separated into several parts (For a detailed compilation of guidelines see McCusker et al. 1999).

Data Collection

It is essential that the powder diffraction data be collected appropriately and with an accuracy as high as possible. This concerns both the instrument and the specimen. Prior to data collection the geometry of the instrument and the quality of the alignment must be checked. The 2Θ scale should be carefully calibrated using several peaks from a standard sample. This is important in Rietveld refinements since the lattice parameters are included in the course of the structure refinement. Radiation and wavelength suitable for the experiment must be selected carefully, for example if the sample contains elements with the wavelength near an absorption edge.

Background is an important feature in the analysis of powder diffraction data by Rietveld full pattern refinement. Special attention should be taken to the background. It should be as low as possible, which can be achieved for example by a special specimen holder, like a $Si(511)$ single crystal, evacuated beam path, a crystal analyzer mounted in the diffraction path between specimen and detector, etc.

Specimen preparation is another most crucial point in every diffraction experiment. Particle size, specimen thickness and "preferred orientation" are common errors. A specimen with inappropriate particle size may ruin the whole experiment and especially the ensuing refinement. Careless packing and hard pressing the specimen into a flat container may result in so-called "preferred orientation", better in a none-statistical distribution of the crystallites. This effect becomes even more serious when using synchrotron radiation because of the highly collimated nature of the incident beam. Rotation of a flat-plate specimen is strongly recommended. It will reduce, but not correct for "preferred orientation" effects. Recommended are particles between 1 and 5 μm, selected by grinding and sieving. Larger crystallites in the specimen give problems, since there the random distribution requirement is not given and there are fewer crystallites diffracting. This can cause non-systematic inaccuracies in the relative intensities. Unlike preferred orientation there is no way to correct for it. Bad is one or several larger crystallites in the specimen, by crystallographers called "rocks in the dust". This may lead to "spikes", affecting only a few, perhaps only one intensity and consequently will give difficulties to the whole analysis or refinement. It will show up in intensity difference plots. Thickness of the specimen is the second source for obtaining poor results. This is described as micro-absorption. Particles too small will result possibly in line broadening, which shows up in the *FWHM* data in comparison to standard samples.

Negligence in the specimen preparation can never be compensated later by the ensuing follow-up steps, like profile fitting, corrections for preferred orientation or in the final least squares refinement. Time and effort spent on the specimen preparation is time well spent.

It is important to select the proper step size $\Delta(2\Theta)$ and counting time per step. This will determine directly the total time needed for the experiment. This of course relates to the purpose of the study and the time available for the experiment, which, for example, may be limited at synchrotron facilities. Step sizes between $\Delta(2\Theta) = 0.01$ and 0.001 are possible without much difficulties. The consequences are extremely long counting times of many hours. One thousandth of a degree yielding 80 000 data points in a pattern of 2Θ up to 80° takes about 44 hours (see Cox et al. 1983). To avoid, or at least reduce background a crystal analyzer mounted in the diffraction path between specimen and detector is recommended.

Considering the step size. This depends strongly on the resolution of the instrument and setup. There should be at least five steps across the top of each peak. Using more than 10 steps is a waste of time, since it will not improve the profile analysis. The best step size is ca. 0.01° to 0.02° in $\Delta(2\Theta)$. If there is a gradual decline of intensity, like in synchrotron experiments, care must be taken of this influence, for example by a primary beam monitor.

The errors most common occurring in diffraction experiments are the following: In Bragg-Brentano, the most common geometry used, it is important to ensure that the incident beam be kept on the sample at all angles to ensure "constant volume" conditions. This is determined by the divergence slits selected, which may be too wide, so the beam hits the sample holder at low angles resulting in intensities measured too low at low angles. Here it helps, if the instrument is equipped with an automatic divergence slit system which opens as a function of diffraction angle. It is advisable to develop a measuring strategy for a successful refinement. Good counting is required and this requires for example to spend more time at high angles where the intensities are low due to the atomic form factors (scattering factors) and to a lesser degree due to the influence of temperature factors.

Another problem is that of specimen transparency. The data are collected under the assumption that (in reflection geometry) the specimen is "infinitely thick", i.e. the X-ray beam is totally absorbed by the specimen. If only light elements are in the sample this condition and the constant volume assumption might not be fulfilled. As a consequence the intensities measured at high angles will be too low. For structure refinement this will be not a serious problem, since this deficiency will be absorbed by the temperature factors, which then of course have no physical meaning any more and should not be used for analysis of thermal motions. In such cases and quite general however transmission geometry, i.e. Debye-Scherrer geometry, is to be preferred.

Handling of Experimental Data

There may be a temptation to smooth the diffraction data before doing profile or structure refinements This definitely must be resisted. Profile fitting will absorb eventual deficiencies in the counting statistics. The refinement will still yield the more or less correct atomic parameters, only the R-indices may not be up to the user's expectation. Smoothing also introduces point-to-point correlations and will give falsely lowered e.s.d. values in the refinement.

Background

Background is a contribution to the data with which the experimenter has to struggle most. Background can never be avoided completely and must be dealt with one way or the other. Of course appropriate experimental provisions are the best way to reduce this problem. One is to evacuate the beam paths, or flush them with helium gas. Another is to use a single crystal specimen holder, for example a silicon plate cut to (511). An analyzing crystal between specimen and detector is another means. The background should be subtracted before the actual refinement. If the background is described by a polynomial and if the background is subtracted it should be re-estimated and re-subtracted several times during the refinement. Also the experimenter must be aware that the background influences the R-values very strongly. The higher the background, the lower the R-index (see Sect. 3.6).

In the description and refinement of background two mathematical approaches are commonly used: The most general one is by a linear interpolation between selected data points between reflections, so-called foot marks. It is preferred because it is simple and is included in almost every Rietveld program version. It is however a good solution only for simple patterns where most peaks are well resolved to the baseline. For more complex patterns with many reflections or low symmetries the majority of the peaks are not resolved to the baseline. In these cases polynomial functions can be used to describe the background. The factors in the polynomial can be included in the refinement. Usually the polynomial functions used for this purpose are largely empirical. In a paper by Lauterjung et al. (1985) the background is described and refined by a sum of n orthogonal weighted polynoms, a procedure based on a method given by Steenstrup (1981). The function used is represented as a Tschebyscheff-like polynomial giving individual weights to each data point in order to distinguish background from signal. With a polynomial of order five extremely good results were obtained.

Peak Shape Function

An accurate description of the shapes of the peaks is critical to the success of a Rietveld refinement. If the peaks are poorly described the refinement will not be satisfactory. The peak shapes observed are a function of both sample and instrument contribution. They may, and often do vary as a function of 2Θ. Accommodation of all aspects in a single-peak description is nontrivial and compromises have to be made. The several peak shape functions are discussed in detail in Sect. 1.2.5.

Regardless of the type of function selected, the range of a peak must be established, meaning when does the peak no longer contribute significant intensities to the diffraction pattern. This is not trivial, since a high Lorentz contribution leads to long "tails". As a rule of thumb a peak is normally considered to be down to background level when the intensity is less than 0.1 to 1.0% of the peak maximum. It can be expressed through the $FWHM$. The range needed will range typically from 10 to 20 times $FWHM$. This means a reflection with a value of $FWHM = 0.10°$ (2Θ) contributes to the diffracted intensity over a range of $1.0°$ to $2.0°$ (2Θ). It is recognized in many publications that this parameter is often set much too low which then shows up in the difference profile plots.

Peak Finding

If a pattern is poorly resolved it may be advisable to determine the number of peaks under the diffraction pattern. This is a "must" in the Two Stage method and for the purpose of *direct methods*. Such an analysis and separation is necessary for example in complex high pressure patterns, often consisting of several unrelated phases. Or in neutron diffraction patterns where satellite peaks from magnetic ordering are close to the main peak and must be found. A peak separation method has been developed by Savitzky and Golay (1964). It has been applied in an overcrowded synchrotron diffraction pattern, where it worked well (Lauterjung et al. 1985). In that special case Gaussian peak shapes were assumed, since the second derivative of a Gaussian distribution will produce a Gaussian distribution again, however with reduced *FWHM*. This means that the second derivative of the spectrum will produce a pattern with increased resolution and the peak positions are given as relative minima in this pattern. Therefor the second derivative of the poorly resolved profile will be calculated by this numerical method. This is repeated iteratively adding a new peak at each step until background level of this section is arrived. After all peaks and their positions have been found, one can go back to profile fitting with the profile shape function most suitable for this pattern.

Another similar way is to separate the peaks interactively in front of the screen by indicating possible peaks with the cursor. This enables the user to add or delete peaks from the cluster (Will and Höffner 1992; Merz et al. 1990) (see also Sect. 3.3.1).

Structural Model

As has been stated several times for a satisfactory Rietveld refinement a structural model including approximate lattice parameters and symmetry must be available. If this model is still incomplete, difference Fourier maps can be used to locate missing atoms, as has been done in several examples shown in the literature and also mentioned in this monograph. In general however refinement of structural parameters should not start too early, especially if not all atoms have been found. Most Rietveld programs today provide the possibility of partitioning overlapping reflections, and with some care and experience this approximation may be treated like a pseudo-single crystal data set, which can then be used to proceed with electron density or difference density maps. This approaches the Two Stage strategy.

It must be realized however that maps calculated from powder diffraction data sets determined by partitioning are more diffuse than from single crystal data sets. This is different in the Two Stage method, where integrated intensities are extracted.

Refining a Structure – the Strategy

With a complete and good structural model and good starting values Rietveld refinement can begin, and in the hands of experienced scientists will be a straight forward procedure, – almost straight forward; and successful. Good starting values are needed

i for the unit cell;
ii the background;
iii the profile parameters; and
iv naturally for approximate atomic positions.

It will not work, if the atoms are placed too far away from their final positions. One problem and pitfall in almost every Rietveld refinement is extensive correlation between parameters. It is advisable to do the refinement in consecutive steps with separate groups of parameters in one bloc at a time. This should certainly be done at the beginning of a refinement. Such a separate refinement may follow for example these steps with several cycles at each step:

1. Scale factor alone, two or three cycles
2. Atomic positional parameters, heavy atoms first
3. Occupancy numbers
4. Background
5. Remaining parameters, except temperature values
6. Thermal displacement, i.e. Debye-Waller factors, should be the last refinement (and here it is advisable to begin with the heavy or heavier atoms)

It must go around in many cycles, a total of 100 cycles is not an exception, depending on the size of the structure. After the first go-through it is not necessary to follow always the same order. Also parameters may be combined occasionally, to look for correlations. Occupancy parameters and thermal parameters need special attention and care. They are highly correlated with one another and should never be refined simultaneously. Occupancy parameters are difficult to refine in any case, and without experience this may result in completely meaningless numbers. Here constraints must be included in the refinement. All least squares program have provisions for subroutines, which should be used, and unfortunately sometimes must be programmed. The chemical composition with corresponding constraints is one such necessity (see for example Nover and Will 1981; Will et al. 1992a,b). Also the Debye-Waller factors should be taken with care, if necessary kept constant at a reasonable value (taken from literature). It is known from experience that Debye-Waller factors tend to absorb a number of deficiencies in the diffraction data. These include difficulties in determining proper background, deficiencies in the profile parameters, in the varying change of peak width due to particle size or strain/stress in the crystallites, chemical bonding contribution and so forth. The refinement may occasionally give negative values, then they should be set to zero or kept constant at a reasonable value taken from the literature.

The structure should be refined to convergence, based on the maximum shift per e.s.d. in the final cycle this should be no more than 0.10.

Number of Observations versus Number of Parameters

In a Rietveld refinement the individual observations at each step are the observations in the core of the least squares algorithm. Consequently the number of observations is very high. However only the number of integrated intensities of the individual reflections can be considered unique observations for the refinement of structural parameters. This is a dangerous pitfall, because this number enters directly the calculated R-values tending to give very low, i.e. very good values implying a successful refinement. Also the e.s.d. are directly affected by these numbers implying very small uncertainties of the parameters.

It is important to have some estimate of the amount of information in the pattern in order to judge how many structural parameters can actually and sensibly be refined. The Rietveld refinement algorithm will allow many more parameters to be refined than the data can actually support. This requires experience. The operator has to use common sense and his crystallographic experience and if necessary intervene. A stepwise refinement of different blocks of parameters as indicated before is advised and will reduce this problem in some way. A ratio observation to parameters should be at least three and preferably five, where the observations are taken as reflections, not steps.

Estimated Standard Deviations, e.s.d.

From a purely statistical point of view, each measurement is an independent observation. Intensities measured at different points on the same peak are simply considered two independent measurements of the intensity of that peak. It is important to emphasize that the e.s.d. calculated assume that the counting statistics are the only source of errors. Systematic errors coming for example from an inadequate background estimate, inadequate peak shape description or simply an insufficiently defined structural model cannot be estimated, however they are significant. Therefore the e.s.d. calculated reflect the precision and not the accuracy of the refined parameters.

R-Values

The quality of a refinement is generally checked by *R*-values. Those number are easy to communicate and to compare. However a difference plot between observed and calculated patterns is the best way to judge the success of a refinement. Of the many formulas used to calculate *R*-values the most common ones used have been listed in Table 1.6 (Sect. 1.2.5).

Some Common Problems

Each structure refinement will present its own problems and require imaginative individual solutions. Some problems are of a more general nature and arise in many cases, especially with inexperienced users. We will not remind the user of possible, and quite frequently encountered simple typing errors in the input file, which is not uncommon. The input data should first be checked carefully if the program does not do what you expect it to do.

Sources of failure often encountered:

i The background is not well fitted, the subtraction is incorrect or the function used is not sufficiently well defined.

ii The peak shape is poorly described. This can be seen for example in a characteristic difference plot of one or a few reflections believed to be measured correctly.

iii A mismatch between peak positions in the calculated and observed patterns. Lattice parameters may be the problem, or more often zero offset or specimen displacement.

iv The relative intensities of (only) a few reflections are too high. Probably the specimen is poorly prepared, likely some crystallite being to large and thus leading to spikes.

v There are small un-indexed peaks in the diffraction pattern, something encountered not infrequently. First it should be verified that these peaks do not arise from the sample holder or other causes or from the experimental setup. If this is not the cause, they may be due to small impurities. Try another sample and specimen. If you cannot eliminate them, leave them out of the refinement.

vi Negative temperature factors. It is often seen, but not a basic problem. Just set the values to certain values or to zero and keep them fixed. There may be many reasons, for example no complete absorption of the primary beam in the specimen, wrong background at high angles due to overcrowded patterns, etc.

vii Refinement does not converge. Many possibilities exist, some very trivial:
- Is the space group or symmetry correct?
- Are the lattice parameters (starting values) correct?
- Is the scale factor correct?
- Is the structural model correct and complete? Are all atoms of the model in the beginning cycle accounted for and placed correctly?
- Are there sudden large shifts in the refined parameters? Mostly due to strong correlations between parameters. Separate the refinement of these parameters in bloc cycles.
- Are there strong correlations between parameters? Try bloc refinement.
- Are the parameters corrected? Try bloc refinements.
- Do you need constraints, for example in the chemical composition or in the temperature factors?

2.2
Special Applications of the Rietveld Method

2.2.1
Quantitative Phase Analysis by the Rietveld Method

Quantitative phase analysis is the measurement of the relative abundance of the constituent crystal phases in the sample, for example minerals. Geologists call it "modal analysis". Quantitative phase analysis based on diffraction methods is the only method which is truly phase sensitive rather than element sensitive. It is the only technique capable of determining phase contents in a sample. With the steadily growing interest in whole pattern fitting methods and Rietveld analysis techniques quantitative phase analysis is one of the fastest growing and most important applications of powder diffraction, in part because of its high potential for industrial analysis. Companies dealing with diffractometers have in the last years spent much efforts, time and money to include the proper software packages into their line of instrumentation.

Phase analysis *per se* is a well established procedure, dating back to the thirties when it was introduced by Hanawalt (1938). Extended applications followed the development of the theory by Alexander and Klug (1948). The method is based on observed intensities in a diffraction pattern, generally measured with X-rays. Preferable are integrated intensities of individual peaks for each phase in the mixture. For many years X-ray quantitative phase analysis was confined to regions of diffraction patterns with resolved lines. Moreover standard calibration mixtures was recommended. Peak clusters should, and today can be separated by profile fitting programs, which goes already into present applications. The original Hanawalt technique relied on three strongest peaks in the pattern, consequently there were severe problems with preferred orientations or incorrect peak intensity data in the Hanawalt files. Nevertheless the Hanawalt technique was applied effectively for many decades. But data collection and even the determination of the phases, which had to be looked up manually, was laborious. Since the 1980s with the advent of faster and smaller computers, PCs, this technique has been improved steadily, moving from the manual search in the Hanawalt files to storage on data banks, on diskettes and lately on discs with the appropriate search-match programs. The search was thereby extended from three to five strongest peaks. Today the International Centre for Diffraction Data, ICDD, collects data and distributes the Powder Diffraction File, PDF. It contains selected d-spacings, recently more and more also full diffraction patterns for use in Rietveld based applications. To include basically the diffraction pattern of any crystalline material known, powder diffraction pattern are even generated from single crystal data, if necessary and if available. The file is constantly updated and contains today (2004) 157 048 entries (133 370 inorganic and 25 609 organic). Of those 92 011 are experimental entries.

The "Hanawalt" procedure is qualitative, or semi-quantitative at its best. For quantitative phase analysis it requires calibration by standards, or adding standards to the sample in question. It is obvious that there were attempts to overcome this limitation and to go for "standardless quantitative analysis of multiphase mixtures", as it was for example put forward by Rius et al. (1987). This was a first attempt, but it was rather impractical, since it required a complete qualitative phase analysis to begin with and sufficient samples with essential differences in the phase composition.

A more direct way was described for the first time, and tested, by Weiss et al. (1983) based on a "Two Step procedure". Experimental intensities are determined from the diffraction pattern, including overlapping diffraction lines, and those were compared to absolute intensities calculated from known crystal structure data. The method has been tested on three examples with two phases, with three and with four phases. In the first sample with quartz and siderite no correction for preferred orientation was included. The results of four different mixtures containing quartz contents between 82 and 7% were very good, except for the last sample. The other tests included minerals with severe cleavage properties, for example mica or calcite, and here the results were quite out of the expected range if no corrections for preferred orientation was included. The results improved when the data were corrected for preferred orientation.

Following the rapid development of powder diffraction in almost every field it was obvious to try to use whole pattern fitting procedures or Rietveld methods also for quantitative phase analysis in order to overcome the drawbacks of the "Hanawalt" technique. The need for calibration with standards is one of the most severe and time consuming requirements. Both ways, Two Stage pattern fitting and full pattern Rietveld methods have the potential to perform standardless quantitative phase analysis. They are far superior, and especially faster than the old technique.

The potentials of such an approach was realized already in 1975 and 1979 (Sonnefeld and Visser 1975; Werner et al. 1979). Both publications are basically a "Two Stage" approach, the approach developed and favoured by Will for crystal structure determination and refinement. By pattern decomposition of the step-scan data Sonnefeld separated the pattern into its component Bragg reflections without reference to the crystal structure, other than the unit cell dimensions. Using the unit cell data the diffraction profiles, peak positions and intensities were calculated. Sonnefeld used these derived peak intensities as variables in a fitting procedure This way provides a convenient means to acquire accurate relative peak intensities, which can then be used to determine calibration curves and with them the constituent phases.

In a similar way Werner et al. (1979) published results of a quantitative analysis of two binary samples, α-Bi_2O_3/KCl and CaF_2/SiO_2. Their analysis was based on Guinier-Hägg X-ray film data. The general Rietveld method is for the refinement of just one crystal structure in the specimen. It has been extended by Werner et al. (1979) to treat a multicomponent system. In this first examples two components were refinable. The quantitative analysis of this mixture was achieved by assuming and refining an overall scale factor together with "occupation" parameters, n_1 for all atoms in one component and n_2 for the second component. The weight percentages ω_1 and ω_2 are related:

$$\omega_1 = \omega_2(Z_1M_1 / Z_2M_2)(n_1 / n_2)^2 \tag{2.10}$$

with M and Z the respective formula weights and formula units/cell.

In a slightly different way Taylor and Pescover used whole pattern techniques for the phase analysis of "real life" samples from zeolite bearing rocks. (Taylor and Pescover 1988), again in a two step procedure. They extracted the (hkl) intensities by profile fitting, and with them they first refined the crystal structures for the structural parameters using the program SHELX. The SHELX outputs $F(hkl)$ on the absolute scale for the theoretical zeolite structure, correct (hkl) multiplicities are attached and the profile scale factor is varied graphically until the calculated patterns matches in intensity with its mate in the observed pattern. All phases in the mixture are removed in this way until the residual pattern is as close to zero as possible. This again is a "Two Stage" approach. Zeolites pose specific problems: The framework is of fixed geometry, however the channel contents can vary with locations. Also there is a severe problem of preferred orientation, which had been reduced in some of their experiments by adding fine aluminum powder. Their approach was to subtract successively the (previously with SHELX) refined patterns from the measured diffraction ones. For this purpose a program had been written which allowed such a successive pattern subtraction. All phases in the mixture were thus removed until the residual pattern was as close to zero as possible. The resulting percentages were correct within a few percent. In summary this way is useful for specific problems, however it is tedious and was not taken up by other researchers.

In general there are two kinds of approach to obtain results of quantitative phase analysis: The Two Stage method, as has been demonstrated by the examples mentioned, or by direct Rietveld multiphase analysis, where the percentage of the constituent phases is calculated through scale factors. The idea to apply a full pattern refinement methods for phase analysis is simple in principle: Determine the material in the specimen by the well established search-match methods, and then, assuming the crystal structures of the components are known, make a least squares full pattern refinement of all components. First individually of each crystalline phase found in the specimen, then of all components simultaneously. The amounts of each sample in the specimen can be calculated through the scale factors. This approach has been proposed by Hill and Howard (1987). In this first attempt they demonstrated the method by measurement on binary mixtures of rutile, corundum, silicon and quartz, collected by neutron diffraction. A specific program, QPDA (for Quantitative Powder Diffraction Analysis), had been written in 1991 by Madsen and Hill (1991). This development evolves from its original purpose of crystal structure refinement to include the determination of phase abundance in polycrystalline mixtures. The idea is actually simple and should be straight forward. However the Rietveld method is not easy to use and may deter powder diffractionists for using the method in nonstructural applications. The simple truth is, that the scale factors obtained in separate, or also combined, crystal structure refinements are related to the masses of the crystalline phases in the mixture. Like with any application of a Rietveld refinement the crystal structure must be known, at least in principle, and this is the case in quantitative phase analysis, following as a rule a qualitative phase analysis. Detail on this method and its use is given in Hill (1991).

The scale factors derived from direct Rietveld analysis of multiphase powder diffraction data are related to the phase composition of the mixture by a simple algorithm involving the product of mass and volume of the unit cell contents of each phase. The method is relatively straight forward. The parameters are in general (simultaneously) refined: Structural parameters, experimental parameters and the weight fractions through the scale factors. In this first application Hill and Howard (1987) proposed a simple relationship between the Rietveld scales in a multiphase mixture and the weights of the components:

$$w_p = S_p(ZMV)_p / \sum_j S_j(ZMV)_j \tag{2.11}$$

where w_p is the relative weight fraction of phase p in a mixture of j phases, S, Z, M and V are respectively the Rietveld scale factors derived from the refinement, Z the number of formula units per unit cell, mass of the formula unit and the unit cell volume.

One strength of this method is that overlapping peaks can be modeled. As a disadvantage the often poorly defined profile shape function does not allow to decompose the peaks unambiguously. On the other hand this is of minor consequences since the scale factors and thus the result of the analysis is little or not at all disturbed. It will show up only in the R-index, which in this case is only a number without bearing. There are a number of publications, which quote standard deviations in the order of 0.1 to 0.4%, as shown for example in the Newsletter of June 2002 of the Commission of Powder Diffraction.

We can assume, that the application is almost exclusively performed in home laboratories with sealed X-ray tubes (or rotating anode tubes). The use in connection with synchrotron radiation makes it easier, because of the strictly monochromatic radiation, but the analysis is the same The experimental technique is based on either Bragg-Brentano geometry or parafocusing Debye-Scherrer geometry. We may refer to a paper by Hill and Madsen (1991). They claim from their results and experience that Debye-Scherrer geometry should be the configuration of choice. In this configuration the specimen, generally in a rotating cylindrical glass tube, preferred orientation is greatly reduced. All present-day neutron powder instruments are based on Debye-Scherrer geometry.

Hill and Howard (1987) give a detailed description of the method. Debye-Scherrer data are collected by neutron diffraction, which, to be fair, is not the general method in laboratory and industrial applications. The great advantages are that the entire diffraction pattern is collected simultaneously, a very important fact, if there are problems of sample stability. Further, if it is used with X-rays, it requires very small specimens, and preferred orientation is greatly reduced. Finally the specimen containment, generally cylindrical glass tubes, allows one to construct environmental chambers which make it much easier to study specimens under extreme conditions, like high or low temperature, or even in high pressure devices.

Program packages for computer supported phase analyses are provided by several companies, like Siemens (now Bruker), Seifert, or Philips, but only on a commercial basis with the ensuing costs. These are for example TOPAS from Bruker, ca. US$13 000, or AutoQuan from Seifert, now GE. This limits the availability. It may be justified, since the method of phase analysis is used on a wide scale routinely especially in industrial laboratories for material analysis, sometimes even controling manufacturing processes. It is not general available in universities, which means it is not generally available for the training of students. In a 1991 compilation of existing programs by Smith and Gorter (1991) we find 18 programs listed as "quantitative analysis". However these are well inferior to the modern company owned programs, which were not available at that time and are not included in this compilation.

Equation 2.11 by Hill does not take into account the relative absorption and particle size of the component phases. This is an important and necessary correction, which has been pointed out already by Brindley (1945). According to the Brindley theory Eq. 2.11 should be given as

$$w_p = S_p (ZMV)_p / (\sum_j S_j (ZMV)_j / \tau_j) \tau_p)$$

(2.12)

where τ_j is the particle absorption factor for phase j. This improved formula has been tested by Taylor and Matulis (1991) with significantly improved results. They analyzed their data using the program AIROQUANT, a commercially available software package for quantitative X-ray powder diffraction analysis. They claim that results without including the Brindley absorption contrast corrections with an input of estimated phase particle radii will yield inaccurate results. However particle size is an important parameter.

An example of a quantitative analysis of minerals by multiphase profile analysis based on the refinement of the complete powder diffraction pattern is given by Tayler (1991) using the program TRASCAL. The program is a full matrix refinement of 14 independent instrumental parameters in a Rietveld refinement procedure. These are: phase scales, asymmetry, preferred orientation using the March equation, linewidths, instrument zero, line shapes and unit cell parameters. The crystal structure data are taken from the crystal structure data bank, which contain some 90 common minerals. Structural parameters are refinable, if felt necessary. Matrix absorption, which may lead to erroneous results, is corrected by a factor $\tau(i)$ following the procedure suggested by Brindley (1945). The sum of the calculated patterns, derived from the data bank, is fitted to the observed pattern. A seven-phase natural bauxite pattern containing 320 independent (hkl) reflections has been refined. The sample was hand-grinded to particle sizes of about 5 µm. The profile R-values range from 0.20 to 0.31 for the several minerals. The phase analysis overall can be considered successful.

Quantitative phase analysis using the Rietveld method is becoming more and more attractive to users of quantitative phase analysis. In a recent publication Peplinski et al. (2004) discusses possible problems in using this method. It may impair the accuracy of the results of a quantitative phase analysis when using the Rietveld method. His investigation is based on a certified reference material, silicon nitride powder, containing two crystalline constituents, the α-phase and the β-phase of silicon nitride. As a conclusion the analysis does not necessarily lead to a mathematical stable solution. The critical points are, not unexpected, that the atomic positions and temperature factors at least of the minor phase(s) should be kept fixed at their accurate values (taken from literature) or should at least be restrained.

Round Robin on Quantitative Phase Analysis

The interest in quantitative phase analysis based on full pattern analysis of complete diffraction patterns is becoming more and more important with wide spread applications in many laboratories already today and certainly in future, however sometimes with conflicting results. Therefore in 1996 the Commission on Powder Diffraction (CPD) undertook a broader comparison of laboratories and methods in the field of quantitative phase abundance determination (QPAD) from diffraction data. The results have been documented in great detail by Madsen et al. (2001) and by Scarlett et al. (2002). The aim of these round robin projects was to document methods and strategies commonly employed and to assess levels of accuracy, precision and lower limits of detection. Four samples were prepared and sent to participating laboratories. Sample 1 was provided in a first study (Madsen et al. 2001). It was a simple three-phase mixture of corundum, fluorite and zincite, a composition as is met in a general sense almost daily in many laboratories engaged in phase analysis. Samples 2, 3 and 4 were added to this project at a later time. The compilations point to certain difficulties which may be encountered from time to time, not without severe consequences on the outcome. They cover a wider range of analytical complexity, such as preferred orientation in Sample 2, an amorphous content in Sample 3, and the influence of micro-absorption in Sample 4.

The first Sample 1 comprised eight variations of a simple three phase mixture: corundum, fluorite and zincite, which was deemed to present little difficulties to most diffraction analysts doing such analyses routinely. It met most likely the common samples which the analytical laboratory will be confronted with. These samples provide minimal preferred orientation, line broadening or micro absorption.

The second round with Samples 2, 3 and 4 addressed problems in the samples which can make the analysis difficult and the results perhaps not very reliable. Forty-nine participants took place in this second study. Partly they collected their own data, and partly they analyzed CPD-provided data (89 analyses out of 288). Most returns were analyzed using laboratory-based X-ray equipment, reflecting the prevalence of this method "at home". Six analyses came from synchrotron data and 17 from neutron diffraction data. Most analyses were performed using Rietveld-based methods, with up to ten different Rietveld software packages employed.

Sample 2 had in addition to the three phases corundum, fluorite, zincite also brucite, $Mg(OH)_2$, crystallizing as flat hexagonal platelets with a strongly anisotropic crystal shape with relatively large dimensions in the ($hk0$) plane and small dimensions in the ($00l$) direction. The flat-like morphology induces strong enhancement of the ($00l$) reflections as a result of specimen packing. This is a situation which is met frequently in day to day analyses. Some participants made corrections for preferred orientation by spherical harmonics. This appears to be the best correction method employed. But also correction by the March-Dollase formula appears to yield better results than the absence of corrections.

Another quite frequent situation is addresses in Sample 3, which contained silica flour as an amorphous content, besides corundum, fluorite, zincite. Amorphous contents are in many samples not an infrequent problem. And this situation is one of the major analysis problems altogether. It is mostly ignored in routine analysis in laboratories and it was ignored in this study in several contributions in spite of the obvious presence to be seen in the diffraction pattern.

Sample 4 contained besides corundum, magnetite and zircon synthetic bauxite, natural granodiorite and a synthetic pharmaceutical mixture, a sample were micro absorption had to be considered. The problem of micro absorption appears to be the biggest physical hindrance to accurate quantitative phase analysis when using X-ray diffraction data. (This problem is virtually absent in neutron diffraction) Depending on the material it may be necessary, for X-ray users, to calculate the mass absorption coefficients for the component phases to determine whether or not micro absorption may become an issue. If it is the particle size must be handled properly, either by grinding or by measurements of the sizes of the particles.

As a conclusion of this study it is obvious that it is of the utmost importance that the preparation of sample and specimen be appropriate for the task at hand. The second important point is the data collection method, including appropriate wavelength.

2.2.2
Texture Analysis Using the Rietveld Method

Texture in polycrystalline samples means the distinct spatial orientation of the individual crystallites within the material. In industrial materials, like in metals or ceramics, the orientation of the grains, or the polycrystalline texture, can have profound effects on the physical properties of the material affecting both strength and possible failure. Texture in rolled sheets is such an example. For geological materials, the texture found in rocks is a consequence of the thermal and deformation geological history. These two most important fields are treated quite separately by the scientists involved, metallurgists or geologists. Metals for example consist of fine grains and are fairly easy to analyze with X-rays. Texture in those cases often must be analyzed fast, sometimes even continuously during the manufacturing process. This analysis is based on X-ray diffraction, where the primary, but also diffracted intensities are high and where measurements are fast. The necessary techniques and instrumentation has been developed by several commercial companies for such uses.

Geological samples on the other hand are hampered by generally larger crystallites, up to the cm range in granites, for example, and they require therefore larger specimens. Since about 10 years neutron diffraction has offered here an attractive way out using specimens up to one cm cube. Because of small absorption of neutrons in matter large specimens pose no problem. The problems here are different. The intensity of the primary neutron beam is low requiring long times for one experiment. This has been overcome by position sensitive detectors with the ensuing sophisticated soft ware. This is discussed in detail later in Sect. 3.9.

Texture analysis by diffraction methods, X-rays or neutrons, is based on rotating the specimen through the full three-dimensional Eulerian space and measuring the intensity at each, and many hundreds specimen positions. In the past this was done with one detector position fixed at the center of one individual diffraction peak for one specific lattice plane *hkl* and rotating the specimen into various positions for all angles in the Eulerian space. Then the detector is moved to the next reflection. This takes time and is efficient if only a few pole figures are required to calculate ODFs with sufficient quality. Also if the diffraction peaks are reasonably strong and well separated, such as in fcc and bcc metals. With high intensity X-ray generators, or even synchrotron radiation, it is not a real problem. Other with neutrons. When in the last years position sensitive detectors covering a considerable range in 2Θ came into use the situation has improved also here.

The method of analysis is the same regardless of X-ray or neutron data. The analysis begins with the calculation of pole figures calculated from each (*hkl*) data set. In most cases the orientation of the crystallites in the sample can be derived from those figures, looked at in different directions. The full orientation of the crystal grains within a polycrystalline object is then commonly described by the three-dimensional orientation distribution $f(g)$, called ODF, which is a mapping of the probability of each of the possible grain orientations with respect to a macroscopic sample orientation. Crystallite orientation in rolled sheets is such as specific property. Traditionally, an ODF is calculated from pole figures.

Texture must not be confused with the general, and comparatively small preferred orientation observed to a small degree in almost all specimens (not samples!) used in powder diffraction. Non-statistical distribution of crystallites would be a better term in that cases. Quite often specimens suffer from preferred orientation as the result of specimen preparation, for example if the samples are pressed into the flat specimen holders in Bragg-Brentano geometry. Formulas have been developed to correct for these effects. These formulas have been implemented very early into all Rietveld refinement programs and also into POWLS. Dollase (1986) has developed a general formula for specimens with considerable preferred orientation of the crystallites in the specimen, for example containing calcites or micas with profound cleavage properties. Dollase's formula is based on a mathematical treatment discussed by March (1932).

Advances in Instrumentation and Technique

With the advent of linear position sensitive detectors diffractometers equipped with such detectors can cover a range of about 20° in 2Θ and record simultaneously at each specimen position whole spectra containing signals from a reasonable number of diffraction peaks, even if coming from different phases in a poly-phase sample. This reduces the experimental time considerably from days to hours (Will et al. 1989) (see also Sect. 3.9). The result is a wealth of data, which must be handled and analyzed. This is only a minor obstacle.

Lately reliable area detectors became available, for example based on the Debye-Scherrer technique with flat films in the forward direction. Here the complete diffraction rings can be measured and the texture can be seen directly. This was applied lately for a texture analysis using synchrotron radiation (Wenk and Grigull 2003). The area detector records the complete, through texture disturbed Debye-Scherrer rings on the flat film. Perhaps even more important, it can of course also be applied for regular crystal structure refinements to overcome the effects of preferred orientation. By integrating the Debye-Scherrer rings data free of preferred orientation can be obtained. The limiting factor in any appliance using flat films is the very limited range of scattering angle, since data are recorded in the forward direction only within a very limited range of 2Θ. Nevertheless this method is valuable especially in connection with high and highest pressure research when using diamond anvil cells.

Advances in Software

Texture measurements result in huge data sets, especially if position sensitive detectors are used. Will and his co-workers (Will 1983; Will et al. 1990c) have within their Two Stage method developed a way of analysis by separating overlapping diffraction peaks, on-line or operator assisted, in the many, up to several hundreds diffraction patterns, for example 600 patterns in one case of a hematite ore, into individual integrated intensities and using them for pole figure and ensuing ODF calculations. This is treated in detail later in Sect. 3.9.

Generally the intensity of the diffraction peaks depends on the crystal structure and on the sample texture. The general Rietveld programs consider only the structure, e.g. the arrangement of the atoms in the unit cell. Obviously it would be a great step forward, if structure and texture could be combined in one calculation and refined simultaneously. This idea of using the Rietveld method, or more correctly Rietveld full pattern refinement also for texture analysis was considered and tried already several years ago. In a first paper by Ferrari et al. (1996), and later by Lutterotti et al. (1997) these authors combined structure and texture analysis in one single program. Another approach came from Von Dreele (1995), who tried to integrate the harmonic method into his existing program package GSAS (Von Dreele 1997). He treated the harmonic coefficients directly as parameters, thereby avoiding the lengthy data analysis of the many individual diffraction data and the calculation of pole figures. In a later program MAUD this was extended by Lutterotti et al. (1997) who have provided the up to now most versatile texture treatment in the Rietveld software MAUD. In this program the ODF can be entered explicitly. Both groups have used their methods with neutron diffraction data, explicitly with time-of-flight data. It has not yet been applied to synchrotron images. Further development is certainly required.

Refinement of the harmonic coefficients as least squares parameters has some intriguing aspects. There is however some doubt concerning the applicability of this procedure because of series termination and ghost effects, particularly in cases of low crystal sample symmetry.

This area of application of full pattern analysis is advancing very fast and it is advised to observe the current literature.

Chapter 3

The Two Stage Method

3.1
Introduction and Background

3.1.1
Concept of the Two Stage Method

The Two Stage method, or integrated intensity method, goes well beyond the Full-Pattern or Rietveld crystal structure refinement. It opens the way to any analysis of diffraction patterns, especially also to *ab initio* analyses of unknown crystal structures, to the analysis by Patterson methods and even by direct methods (Giacovazzo and his co-workers: Altomare et al. 1996a,b; Carrozzini et al. 1997; Cascarano et al. 1992). It is a very powerful method for structure refinements in connection with the computer program POWLS (Will 1979; Will et al. 1980b, 1983a,b), where the method shows its power especially if unconventional parameters have to be included in the model calculations. One more important application is in using it for texture analysis.

The Two Stage method, as the name implies, operates in two steps, which are completely independent of each other. The first and most important step is fully independent of any crystallographic goal. It is not even limited to crystallography. Any diagram exhibiting peaks can be analyzed by the programs used in Step 1. Table 3.1 gives an outline of the method.

Both methods, the Rietveld method as well as the Two Stage Method in common is the requirement to have a good mathematical description of the profile and the success and quality of the ensuing calculations depend on the validity and reliability of the profile function. *Profile analysis* or *peak-shape analysis* must therefore precede any further calculations. Once the profile function is known (Step 1a), *profile fitting* is performed (Step 1b) and this results in pattern decomposition in the Two Stage method.

Figure 3.1 gives the diffraction pattern of quartz collected with synchrotron radiation at a wavelength of 1.0 Å out to $2\Theta = 133^\circ$. The step width was $\Delta(2\Theta) = 0.02^\circ$, the time spent per step was $t = 4$ s. In most cases dealt with today, including the example of quartz, powder diffraction diagrams have many peaks superimposed onto each other and consequently the diagrams are heavily overcrowded. It is the task of the experimenter, and the purpose of this article, to separate these peaks into its individual contributions. If this is accomplished, powder diffraction is competitive with single crystal diffraction, in many cases it is even superior.

The Two Stage method was developed independently of the efforts by Rietveld. In the beginning it was, just as Rietveld did it, only for the refinement of crystal structures from individually extracted intensities, from X-ray (Will) or neutron diffraction

Table 3.1. Schematic outline of the Two Stage method

STEP 1a	Peak shape analysis (profile analysis)	
STEP 1b	Pattern decomposition (profile fitting)	
STEP 2	Crystallographic computing	
	a Least squares analysis for	
		i Lattice parameters
		ii Crystal structure parameters (like x, y, z etc.)
	b Structure analysis	
		i Fourier calculations
		Patterson maps
		Fourier maps (electron density)
		Difference Fourier maps
		ii Direct methods for solving unknown structures
	c Other calculations	
		i Chemical bonding through HO (high order) and LO (low order)
		ii Anomalous dispersion
		iii Particle size analysis
		iv Strain/stress analysis
		v Phase analysis of unknown composition
		vi Texture analysis

data (Rietveld). For the first application and the first refinement the program POWLS (Will 1979) was developed and used in the analysis if the crystal structure of $MnSO_4$ (Will et al. 1969). Profile analysis and profile fitting was first published in 1987 (Will et al. 1987a,b).

When the Rietveld method became generally known, several scientists expressed concerns of the outcome of the method. They realized the shortcomings of a total pattern refinement, where each count $y(i)$ at each step $\Delta(2\Theta)$ is treated as an independent observation. The Rietveld full pattern refinement method is doubtless very powerful and successful, but statistically the method is unsound and gives incorrect values for the estimated standard deviations, e.s.d., of the structural parameters, even though the values themselves may be perfectly acceptable. This was pointed out by Cooper and Sakata (Sakata and Cooper 1979; Cooper et al. 1980; Cooper 1983). Unfortunately it is reflected also in low R-values, which imply good results, since R-values are often taken as a norm for the quality of the analysis (see Sect. 3.6; Jansen E. et al. 1994a).

Independently of Will Cooper and his co-workers developed an alternative approach, a new method and a new computer program, called SCRAP (Cooper et al. 1981). They followed a two-step procedure: analysis of the diffraction profile to provide estimates for integrated intensities of the Bragg reflections, followed by conventional least squares refinement based on these estimates. The method is described in detail with examples and a comparison with Rietveld results are given.

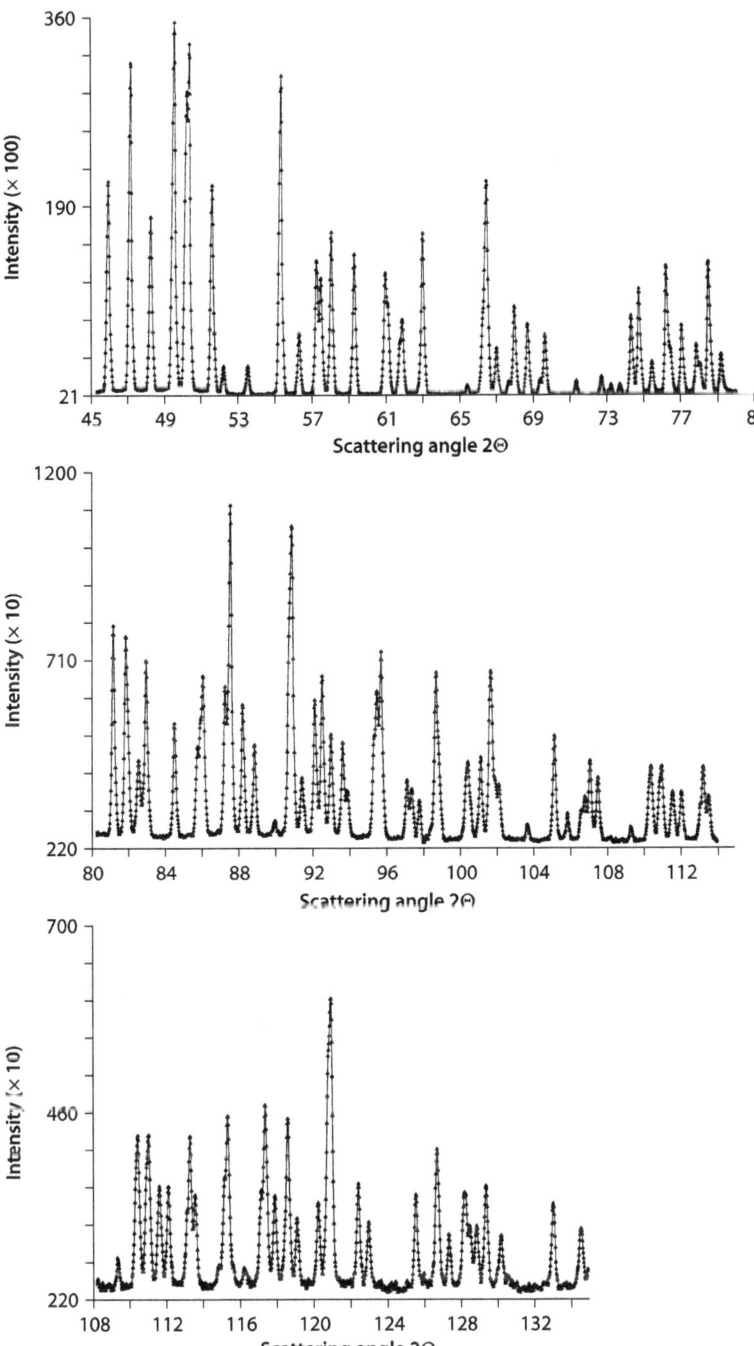

Fig. 3.1. Diffraction diagram of quartz, measured with synchrotron radiation with a wavelength of $\lambda = 1$ Å. The intensity scales of the *lower patterns* were decreased to show the weaker reflections

3.1.2
The Two Stage Method in Comparison to Rietveld Refinement

These two methods commonly used for the analysis of powder diffraction data are not in competition, they rather follow different goals in crystallography. The crystallographer has to decide which method is best suited for his scientific problem. Both have its origin in neutron diffraction: Georg Will 1963 in the Brookhaven National Laboratory, Long Island, N.Y., USA (Will 1979) (in cooperation at the beginning with Walter Hamilton) and Hugo Rietveld at the Netherlands Reactor Centrum in Petten, The Netherlands (Rietveld 1967, 1969). Rietveld's aim was to refine a (neutron) diffraction powder pattern in one step based on a well established structural model (it was extended much later to the application of X-ray data). The history and development has been discussed in Sect. 1.2.

The major difference between the Rietveld method and the Two Stage method is that in the latter we first determine integrated intensities of the individual reflections from the measured pattern. Then the structural parameters are refined or handled with other programs and methods in a manner analogous to single crystal structure determination, as listed in Table 3.1, Step 2. There are much more possibilities than just structure refinement. In the Rietveld method, both the measured intensities and the structural parameters are simultaneously refined using the whole pattern. It is a refinement of a (believed) known crystal structure model. The Rietveld method consequently is a full pattern analysis (Rietveld 1967, 1969). The Two Stage method was originally also developed in connection with neutron diffraction as early as 1962 (Will et al. 1965) although the integrated intensities at that time were derived by conventional methods. Originally the Two Stage method was applied to refine crystal structures from diffrac-

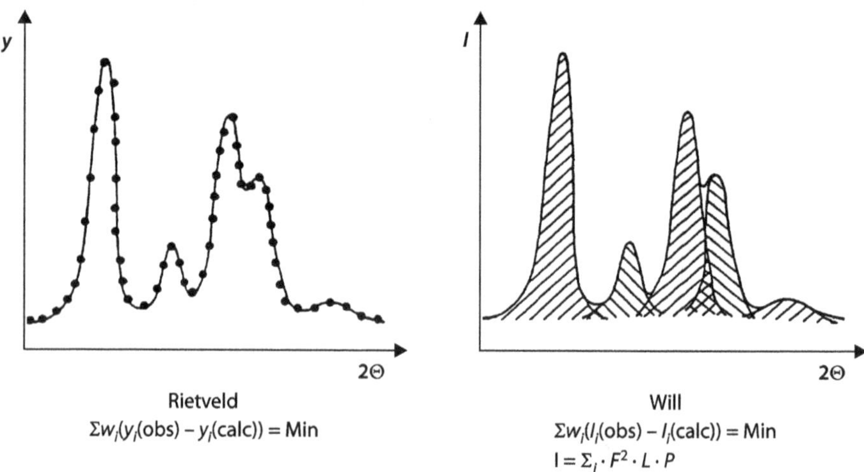

Fig. 3.2. Schematic representation of the two analytical methods, the Rietveld method and the Two Step method. In the former each measured point $y(2\Theta)$ is used as an observation in the least squares structure refinement. In the latter the peak clusters are separated, and those integrated intensity data $I(hkl)$ are the observations for ensuing calculations, just as with single crystal data

tion data collected both with X-rays and with neutrons. It is based on a general least squares program POWLS (POWder Least Squares) adapted for the refinement of crystal (and magnetic) structures from X-ray as well as neutron diffraction data (Will 1979, 1989; Will et al. 1983a).

In later developments both methods were extended also to a priori crystal structure determinations. The Rietveld method, as all full pattern methods have its limitations because they do not unscramble the diffraction patterns in its components of individual integrated intensities for each reflection. If intensities are necessary they begin with the peak heights as a reasonable information on the intensity. In the ensuing refinement of the crystal structure the Rietveld program takes into account the profiles of the diffraction peak.

Figure 3.2 shows schematically the two methods. Both are based on a least squares formalism, whereby the Rietveld method takes the individually measured values of each step in a step-scanning diffraction experiment $y(i)$ $(= y(2\Theta))$ as the observations, while the Two Stage method first separates in its first step of the analysis peak clusters in the diagram into its individual components, e.g. individual peaks, yielding integrated intensity values $I(hkl)$ for each reflection. These are the observations and they are then used for further crystallographic calculations.

3.2
Profile Analysis and Profile Fitting

3.2.1
Characteristics of Powder Diffraction Line Profiles

The observed diffraction pattern originates from the diffraction effects from the specimen. The specimen contribution can be idealized by a delta function convoluted with functions describing line broadening by

- geometrical and instrumental aberrations
- wave length spread in case of monochromatized radiation, especially with X-rays from X-ray tubes with its $K\alpha_1/K\alpha_2$ contributions
- line broadening by particle size effects and/or strain and stress contributions

The observed experimental contribution is therefore a convolution of all effects, and this is superimposed, i.e. added to the background scattering. The main characteristics of diffraction lines are position, maximum intensity, area (integrated intensity) and shape. The latter is characterized by width and asymmetry. Diffraction patterns are interpreted to a particular level of approximation and accuracy using:

a line position: 2Θ location of the maximum intensity, for lattice parameter determination

b range of peak, i.e. the 2Θ region where the intensity is assumed to pertain to the diffraction effect

c peak height: intensity above background at the 2Θ position of the maximum intensity

d integral intensity: area of the diffraction line above background e line width: *FWHM*, full width at half maximum, i.e. the line breadth at half the peak height; or sometimes integral breadth β, i.e. the width of the rectangle having the same area and height as the observed line profile

f profile descriptions using shape parameters: analytical functions, like Gaussian, Lorentzian, Voigt, Pseudo-Voigt, Pearson VII, or split versions of these functions

g asymmetry parameters: asymmetric intensity distributions can be defined by separate breadth and shape parameters on each side of the peak maximum, or in terms of the ratio of parameters between the sides

3.2.2
Outline of Profile Fitting

The crystallographic information of an individual Bragg reflection is contained in the total integrated intensity which is spread over a characteristic profile. This must be extracted from the pattern. The analytical approach is by profile fitting. This is performed in three steps:

- Subtraction of the background
- An analytical description of the instrument-dependent profile shape of the measured Bragg reflections (this needs calibration of the instrument by standard samples)
- Separation of the individual reflections hkl and determination of the individual reflection parameters and integrated intensities

The integrated Bragg intensity is finally extracted by profile fitting. This is performed by least squares calculations with a (given and known) profile function by minimizing

$$\phi = \sum_i w_i \left(y_i(obs) - y_i(calc) \right)^2 = \text{Minimum} \tag{3.1}$$

with y_i the observed or calculated intensity at each 2Θ-point. The summation index i is running over all points in the diffraction pattern. Sometimes at special instrumental conditions it is preferred to use routines where the summation includes just one peak, a peak cluster, or a section of the pattern selected for fitting. In this specific case the range of the segment to be fitted is adjustable interactively on-line by the operator. Typical the number of points i is in the order of 4 000 to 10 000, if the whole diagram is to be fitted, for example with the program FULFIT (Jansen E. et al. 1988) or with a Rietveld program. This assumes for example 40 to 100 point per degree at a step size of $\Delta(2\Theta) = 0.025°$ or 0.01°, respectively, representing $2\Theta = 10$ to 110°. Such values are common. If the pattern is segmented into sections, for example when using the fitting program HFIT (Höffner and Will 1991), those numbers are much less. w_i are the weights given to each observation. They are taken from the experimental error margins σ_i, which are assumed to be proportional to the square root of the count rate $y_i(obs)$ following Poisson counting statistics,

$$w_i = 1/\sigma_i^2 \tag{3.2}$$

Full pattern fitting, like FULFIT and also some other programs (Pawley 1981; Rodriguez-Carvajal 2001; Toraya 1986), stop just one step before the Rietveld crystal structure refinement. Linearisation in the least squares program is reached by a Taylor series expansion. The goodness of the fitting is expressed by R-values.

This profile fitting ends with a list of Miller indices (hkl) and its corresponding integrated individual intensities, the peak positions 2Θ and the $FWHM$ values. The profile fitting method is inherently capable of determining integrated intensities and also the reflection angles with high accuracy because

i   it uses the experimentally determined instrument function
ii   it can resolve overlapping reflections with far greater resolution than the diffractometer
iii   it uses all the measured data points and
iv   it works over a very large range of intensities

3.2.3
Profile Functions

A profile function must be known and presented to any profile fitting program. In the original Rietveld program (Rietveld 1967), developed for neutron diffraction patterns, a Gaussian function was used with reasonably good results and relatively low R_{Bragg}-values. Gaussian is also a good approximation for energy-dispersive diffraction profiles. When the methods were extended later to the analysis of X-ray diffraction patterns collected in laboratories with $K\alpha_1/\alpha_2$ lines a number of authors have recognized that difficulties occur when a Gaussian function was used because of an inaccurate definition of the profile shapes. In the years following other types of functions have been applied with varying degrees of success. Functions commonly used, or at least tested, are Gaussian G, sometimes even double Gaussian GG (Fig. 3.5) (Will et al. 1987a), Lorentzian L, Pseudo-Voigt, Pearson VII or a Voigt function itself, which is a convolution of a Gauss and a Lorentz function. A Gaussian function describes the shape of the main peak very good at the top, and the Lorentz function gives the correct asymptotic behavior in the tails of the reflection. This for example demonstrates the importance of determination and subtraction of the background before profile fitting. The intensity in the tails of the reflection was found to affect the thermal parameters substantially. The thermal parameters from a Gaussian refinement are systematically too large. Gaussian G and Lorentzian L are described in Eqs. 3.3 and 3.4:

$$G(2\Theta) = h\left(-4\ln 2\left((2\Theta - 2\Theta_0)/FWHM\right)\right)^2 \tag{3.3}$$

$$L(2\Theta) = h\left(1 + 4\left((2\Theta - 2\Theta_0)/FWHM\right)^2\right)^{-1} \tag{3.4}$$

The Voigt function is the best function to describe the experimental profiles, it is however difficult to incorporate in profile fitting programs and this function is not found in programs normally in use today. Good results are obtained by a so-called Pseudo-Voigt function, PSV, which is an additive superposition of a Gaussian and a

Lorentzian function with a fractional parameter η relating the ratio of the peak intensities of G and L:

$$PSV = G + \eta L \tag{3.5}$$

The value η must be determined before profile fitting (or Rietveld analysis) can begin. In the Two Stage method this will be Step 1a. It can be done with well developed reflections from standard samples. LaB_4 has been recommended recently. Data coming from laboratory X-ray sources show in addition strong asymmetry (besides the problem of the α_1/α_2 splitting). As it turned out the best description finally found is a split Pseudo-Voigt function, where the right and left halves of the reflections are treated separately (see Fig. 3.3). In neutron diffraction Howard (1982) approximated the asymmetry of peaks by sums of Gaussians.

The quality of different profile functions is demonstrated in Fig. 3.3. where the (111) reflection of silicon has been fitted with the different functions. Looking at the intensity difference plots in the middle of the diagrams the shortcomings of using a single function can be clearly seen: The Gaussian function describes the peak well in the maximum, however the tails are poorly fitted giving a profile R-value of only $R = 9.7\%$. A Lorentzian function describes the tails better, but now the peak maximum is too low. The profile R-value still high with $R = 9.2\%$. The Pearson VII function describes the profile overall better than a single Gaussian or a single Lorentzian, but still the result with $R = 5.8\%$ is not satisfactory. Finally with a Pseudo-Voigt function an acceptable good result is obtained with $R = 2.0\%$. Since the diffraction peaks are mostly asymmetric, even with synchrotron radiation, the best result is finally obtained with a split Pseudo-Voigt function where the right and left side of the maximum are treated with separate *FWHM* parameters giving very good fitting with $R = 1.4\%$. Table 3.2 summarizes the results obtained with the different functions. But it must be kept in mind

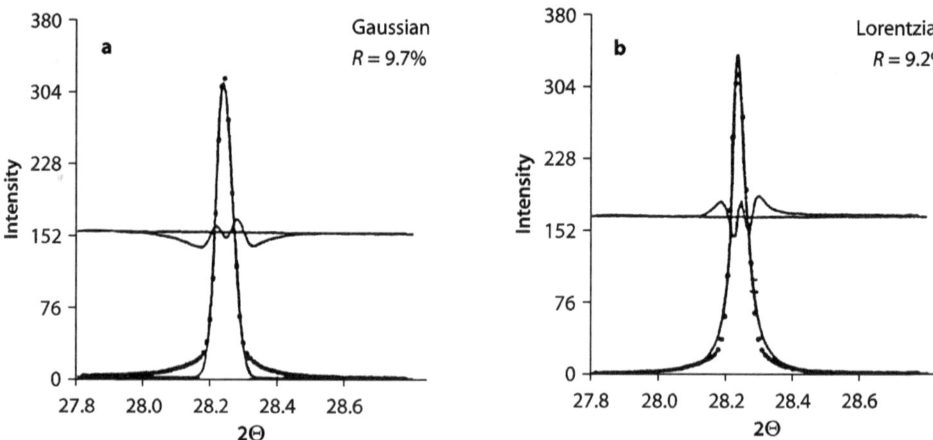

Fig. 3.3 a,b. Measured (111) reflection of silicon fitted with different functions: (**a**) with a single Gaussian function the peak is well described in the maximum, however the tails are poorly fitted. The profile R-value $R = 9.7\%$ is very poor. In (**b**) only a Lorentzian function is used, again the fitting is not good

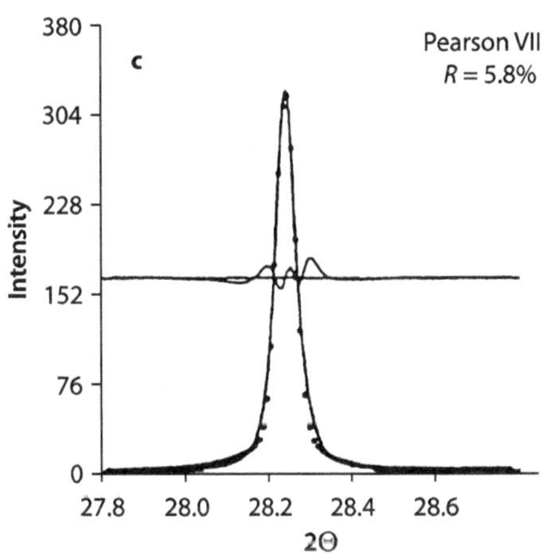

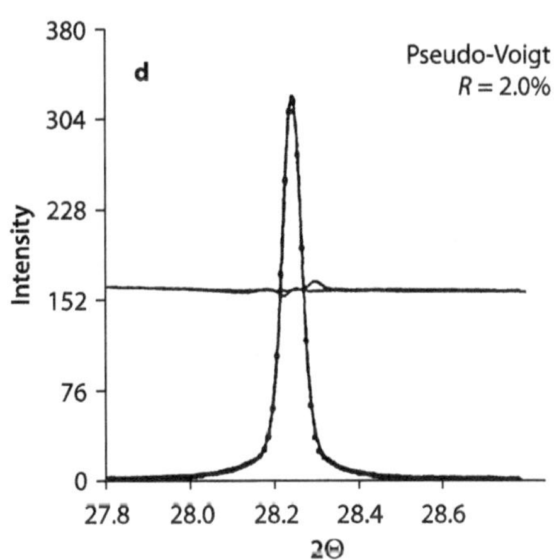

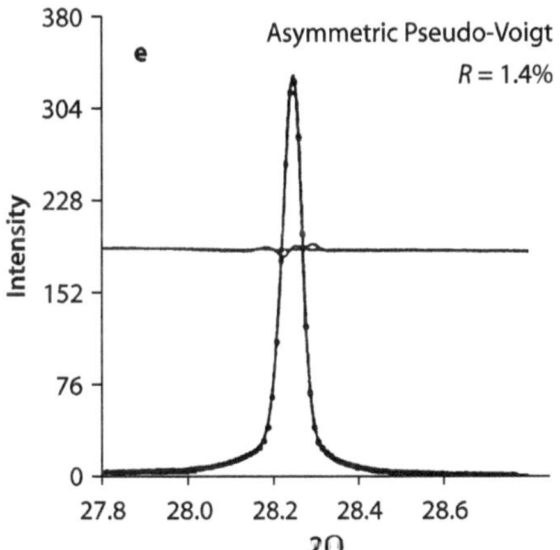

Fig. 3.3 c–e. Measured (111) reflection of silicon fitted with different functions: (**c**) shows the result with a Pearson VII function, which describes the profile better, but still not satisfactorily. Finally in (**d**) we are using a Pseudo-Voigt function with Gaussian and Lorentzian contributions and get a good fit with $R = 2.0\%$. Since the diffraction peaks are almost always asymmetric, in (**e**) finally a split Pseudo-Voigt function gives with $R = 1.4\%$ the best result

Table 3.2. Comparison between different profile functions

	Gaussian	Pearson VII	Pseudo-Voigt symmetric	Pseudo-Voigt asymmetric	Numerical integration
R (%)	9.70	5.80	2.00	1.47	
FWHM left	0.0598	0.0584	0.0602	0.0617	
FWHM right	0.0598	0.0584	0.0602	0.0578	
Integrated intensity	2056.0	2443.8	2510.2	2492.8	2442.4

that occasionally neither of these known functions will describe the peak profiles satisfactory.

The profile function chosen therefore needs special attention in profile fitting. The Pseudo-Voigt function turned out to be the best profile function in most application. In this case only the parameter η has to be determined. In our own experience the profiles were best described by about 85% Gaussian and 15% Lorentzian.

The profiles depend highly on the receiving slits used. In our experiments we normally used a narrow receiving slit. Nevertheless sometimes profiles are found which do not follow the usual profiles. In a set of synchrotron experiments using a crystal analyzer they could not be described even by a Pseudo-Voigt function, or any other known function. A Gaussian curve gave systematically too low peak intensities, higher tails and about 2% too low integrated intensities. A satisfactorily description of the profiles was finally achieved by superimposing two Gaussian functions $G1$ and $G2$ with the ratio 2:1 for $FWHM$ and peak height. This "double Gaussian" $GG = G1 + G2$ (Eq. 3.6) is mathematically simple and makes it easy to handle complex overlapping peak clusters.

$$GG(2\Theta) = 2/3\,h\,(G1(2\Theta) + 1/2\,G2(2\Theta)) \tag{3.6}$$

This profile was found necessary, when we investigated Y_2O_3 (Will et al. 1987a). It was observed in all synchrotron diffraction patterns in this duty cycle (using a crystal analyzer). So this must be due to some unaccounted experimental deviations from normal. Figure 3.4 shows this for silicon (511/333) as an example. The profiles in Y_2O_3 fitted with a double Gaussian function are shown in Fig. 3.5 for a section of the diffraction pattern.

A different situation was encountered with $MnCrInS_4$ (Will et al. 1990b). Here the profiles could finally be described by the sum of a Gaussian and a Lorentzian contribution, but with the latter shifted by 0.03° to the low 2Θ side. This improved the fitting considerably. In this special (and admittedly strange) case no explanation could be found for this behavior. In both cases the profile R-values were in the range 1–2%. Two profiles are depicted in Fig. 3.6 for the reflections (115) and (157)/(266). These two cases exemplify the importance to have a careful profile analysis (Step 1a) before the actual pattern fitting and refinement.

3.2.4
The Instrument Function

The measured intensity $y(2\Theta)$ of a reflection is the convolution of three contributions:

$$y(2\Theta) = WGS + \text{background} \tag{3.7}$$

W is the intensity distribution of the wavelength spectral line of the primary beam, for example the Cu $K\alpha_1/\alpha_2$ line when using sealed X-ray tubes, G represents all instrumental and geometrical factors which contribute to the shape of $y(2\Theta)$, such as the incident and diffracted beam apertures (slit sizes), alignment and related factors. S is the actual contribution form the sample, the quantity we are interested in. In Eq. 3.7 S

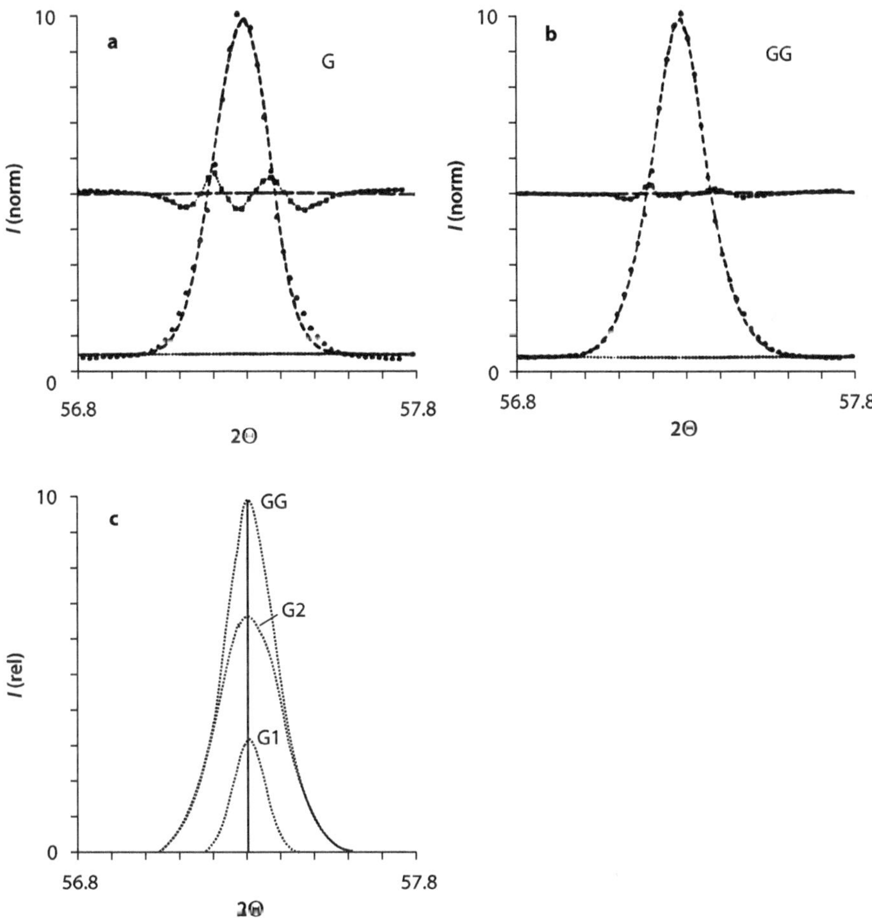

Fig. 3.4. The reflection (511/333) of silicon fitted; **a** with a single Gaussian G function; **b** with a double Gaussian GG function. The differences between experimental points and the fitted profile are shown at half-height; **c** the two Gaussian functions $G1$ and $G2$ and their sum GG. Height and $FWHM$ of $G2$ are twice those of $G1$

is a delta function and can be represented by one symmetrical Lorentzian for each hkl reflection, if the specimen itself produces no asymmetrical broadening, like particle size line broadening or strain/stress conditions. $W \cdot G$ is called the instrument function, which in practice can be determined and treated as one single parameter. Since it relates to the particular experimental conditions, it changes with changing the experimental setup, but stays the same as long as the instrumental setup is not changed. Each new set of slit sizes, wavelengths or other factors contributing to the line shape requires a separate set of $W \cdot G$ profiles and therefore a new calibration. $W \cdot G$ can be determined from standard samples, preferable silicon or other well behaving materials. LaB_4 has been recommended recently and is favored for line shape analysis.

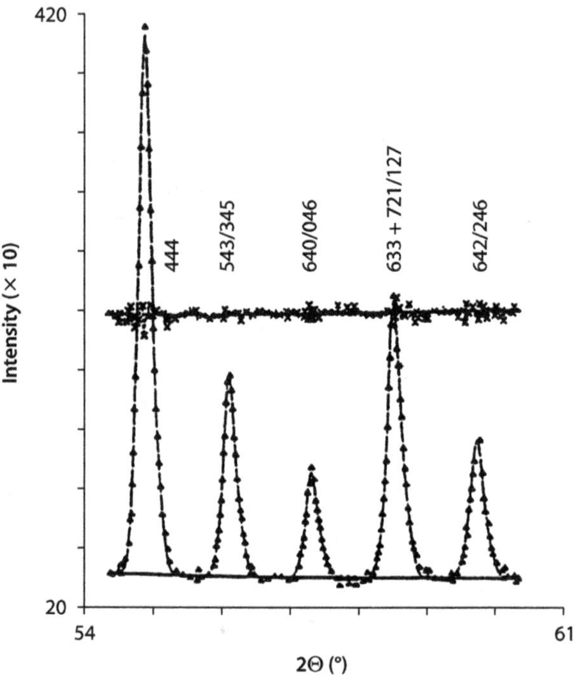

Fig. 3.5. Profile fitting with a double Gaussian function *GG* of five weak reflections of Y_2O_3. $R_{PF} = 1.2\%$. The observed-calculated differences are shown at one-half height

3.2.5
Instrument Function with Seven Lorentzians

When Parrish and co-workers approached the problem of peak analysis and profile fitting of a X-ray diffraction pattern they described each reflection with a sum of seven Lorentzian curves, three each for $K\alpha_1$ and $K\alpha_2$ and one for the weak satellite $K\alpha_3$ (Huang and Parrish 1975). This approach is based on a mathematical description developed by Taupin (1973). It seems a formidable way to describe just one reflections, but it is extremely accurate. It is still the best description of a diffraction profile, however it has not found general acceptance and it is not used today any ore. With position, peak height and *FWHM* for each curve it requires adjusting 21 parameters for each of the reflections used to establish the standard data. Figure 3.7 shows the (331) reflection of silicon, measured with Cu $K\alpha$ radiation and fitted with seven Lorentzian functions.

Why such a difficult procedure? The profiles obtained with powder diffractometers have asymmetries and widths which vary with diffraction angle (Wilson 1963; Parrish 1965). They used the term $W \cdot G$ for the first time to describe the contributions not coming from the specimen. The shapes of the low-angle profiles arise mainly from instrumental aberrations. The high-angle shapes are dominated by the forms of the spectral lines. This shows that it is not possible to have a single instrument function $W \cdot G$ for the whole diffraction pattern. The $W \cdot G$ function is independent of the specimens to be studied. It is therefore essential that the W·G function is determined over

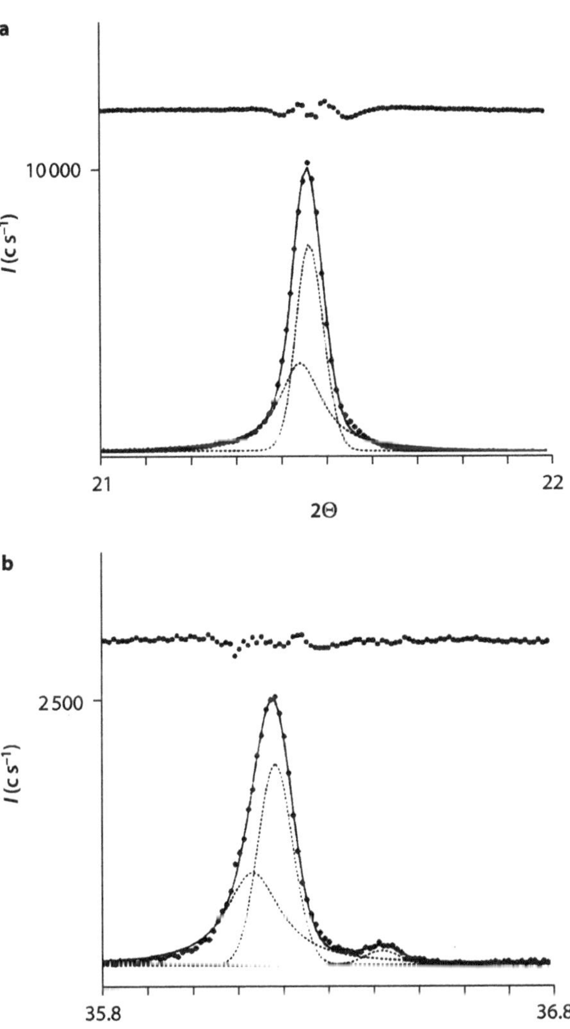

Fig. 3.6. Two representative reflections of $MnCrInS_4$ fitted (*dashed lines*) by a Gaussian and a Lorentzian contribution, separated by a small amount of 0.03° in 2Θ; **a** for the (115) peak; **b** for (157) and (266). The small peak (266) was fitted by only one curve. Differences at the *top*

a large angular 2Θ range, and furtheron that the specimens selected as standards must be well crystallized and must be free of line broadening. Parrish and co-workers placed special attention on the selection of the standards and on the 2Θ dependent instrument functions. Even if this procedure has been given up today, it is worth discussing it. From the samples: silicon, tungsten, α-quartz, diamond, gadolinium-gallium garnet (GGG) and zeolite specimens were carefully prepared by grinding and sifting. To cover the angular range 20° to 110° they measured about 25 profiles with a 1° divergence slit. For the high angle region 80–160° they used a dozen profiles with a 4° slit. The peaks were measured with step sizes $\Delta(2\Theta) = 0.01$ or 0.02°. Counting times were long enough to accumulate about 50 000 counts on the $K\alpha_1$ peaks (Parrish and Huang 1980). Four typical profiles are shown in Fig. 3.8. Differences between experimental

Fig. 3.7. Fitting of the measured (331) reflection of silicon with 7 Lorentzian functions

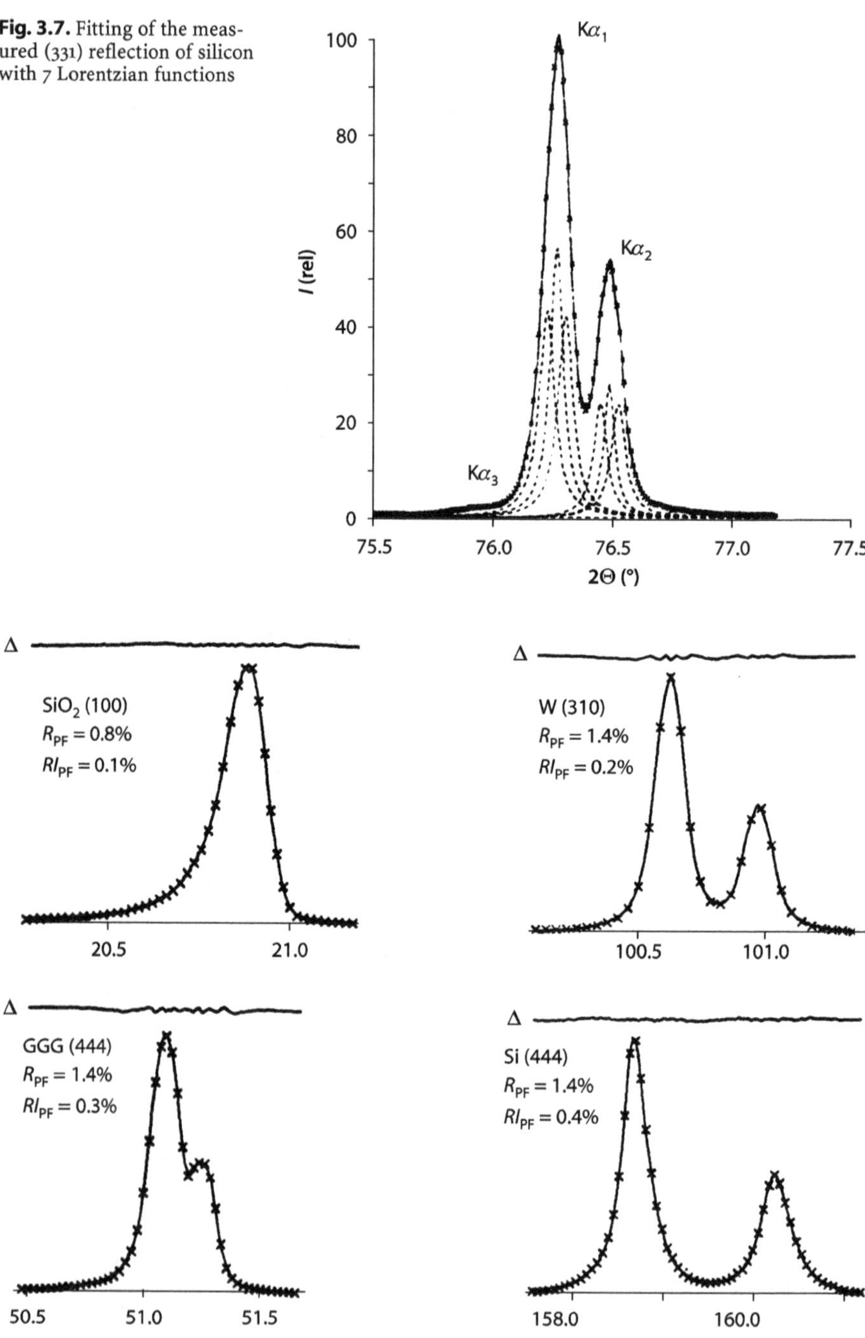

SiO₂ (100), $R_{PF} = 0.8\%$, $RI_{PF} = 0.1\%$

W (310), $R_{PF} = 1.4\%$, $RI_{PF} = 0.2\%$

GGG (444), $R_{PF} = 1.4\%$, $RI_{PF} = 0.3\%$

Si (444), $R_{PF} = 1.4\%$, $RI_{PF} = 0.4\%$

Fig. 3.8. Four typical profiles to determine the instrument function $W \cdot G$. Seven Lorentz functions are used to describe one profile. Alternate experimental points x are omitted for clarity. *Solid curves* are computed from profile fitting. Horizontal Δ-lines above the profiles show the differences between experimental and calculated points on the same intensity scale

and fitted data are extremely small. The goodness of fit is expressed by profile R-values. This procedure yielded the best possible fitting ever. The principle of the profile fitting of a diffraction peak and extraction of the specimen contribution is shown in Fig. 3.9.

3.2.6
Determination of Integrated Intensities through Profile Fitting

The fitting of the reflections in the final step to determine integrated intensities of the specimen from the experimental pattern is based on Eq. 3.8:

$$P(2\Theta) - (W \cdot G) S(I_p, FWHM, 2\Theta_p) + \text{background} \tag{3.8}$$

It requires only the specimen contribution S, which can be fitted with only three parameters: position $2\Theta_p$, height h, and $FWHM$. This is shown schematically in Fig. 3.10. The (integrated) intensities are the values S in Eq. 3.8. The specimen contribution S is obtained by shifting the standard profile over the experimental profile and adjusting the three parameters 2Θ, height, and $FWHM$ (which together give the integrated intensities) by least squares calculations. In the Two Stage method this profile fitting procedure results directly in intensity values $I(hkl)$ together with the 2Θ values for each reflection. The background should have been subtracted first from the experimental data, so it will be zero or at least very small in the calculation. This holds independent of any program used for fitting calculations. Background in the fitting will affect greatly the resulting reliability values, i.e. the R-values, as discussed in Sect. 3.6 and Table 3.9. It will also influence the intensity values and especially their e.s.d.

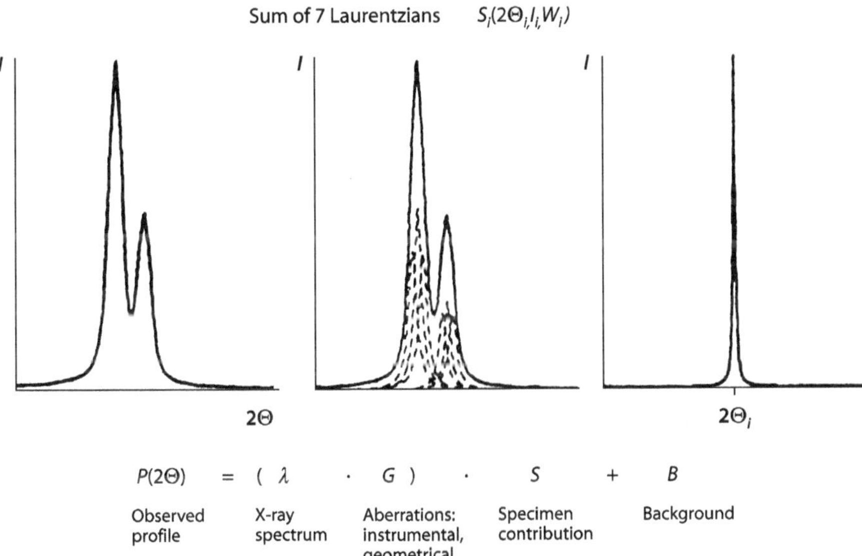

Fig. 3.9. Principle of the profile fitting of a diffraction peak and extraction of the specimen contribution S

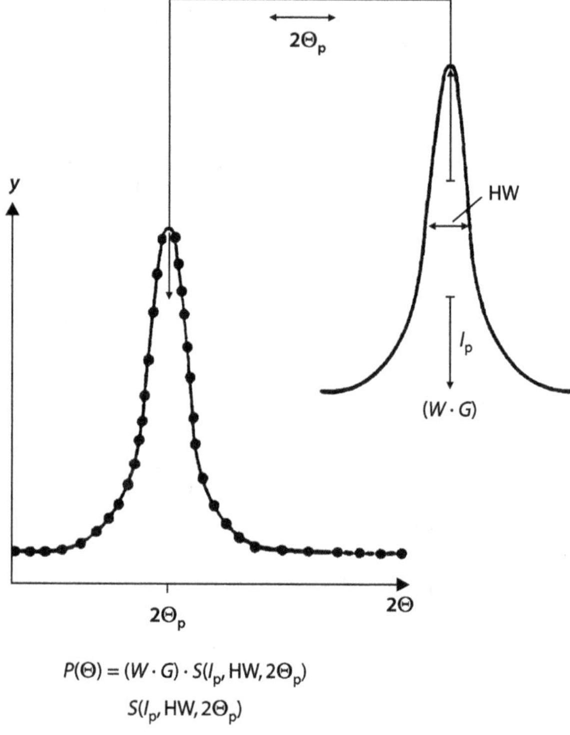

Fig. 3.10. Schematics of the profile fitting procedure: The (known, i.e. beforehand determined) standard profile is shifted over the experimental profile and the three parameters position 2Θ, height h and *FWHM* are adjusted by a least squares calculation

$$P(\Theta) = (W \cdot G) \cdot S(I_p, \mathrm{HW}, 2\Theta_p)$$

$$S(I_p, \mathrm{HW}, 2\Theta_p)$$

The application of the profile fitting does not require prior knowledge of the positions or intensities of the peaks, or any other crystallographic information. In the profile fitting process we minimize, with w_i the weights given to each observation, the equation

$$\sum_i w_i \left(y_i(\mathrm{obs}) - y_i(\mathrm{calc}) \right)^2 = \mathrm{Minimum} \tag{3.9}$$

Integrated intensities can be determined to a high degree of range and accuracy. Tables 3.3 and 3.4 give lists of the resulting intensities for corundum and quartz measured with high precision Cu Kα, using an evacuated beam path and a monochromator, and fitted with the Parrish procedure using seven Lorentzians. The intensity ranges are worth noting. The weakest intensity in quartz is only 28 counts vs. 141 102 counts for the strongest reflection, and in Al_2O_3 we measure 115 vs. 36 531 counts.

3.2.7
Angle Dependence of the Profile Shapes

It is well known that halfwidth and shape of the reflections are dependent on the diffraction angle 2Θ. This has been discussed in detail by Wilson (1963). This variation

Table 3.3. Comparison of observed and calculated intensities for corundum, $\alpha\text{-Al}_2\text{O}_3$, measured with Cu K$\alpha$ radiation

hkl	I(obs)	I(calc)	hkl	I(obs)	I(calc)
012	21535	21505	300	19920	19989
104	33064	33040	125	504	412
110	14082	14062	208	610	571
006	236	288	1.0.10	6468	6133
113	36531	36424	119	2909	2904
202	400	482	217,220	2444	2597
024	17936	17896	036,306	310	332
116	34389	34601	223	1860	1664
211	860	862	131	115	92
122	1210	1142	312	1650	1301
018	2400	2831	128	1000	1028
214	13700	13728	0.2.10	3299	2733

Table 3.4. Comparison of observed and calculated intensities for α-quartz, $\alpha\text{-SiO}_2$, measured with Cu Kα radiation

hkl	I(obs)	I(calc)	hkl	I(obs)	I(calc)
100	29315	29307	220	2229	2272
100,011	141102	141114	213,123	4282	4203
110	10338	10343	221	954	1079
102,012	9267	9250	114	3350	3458
111	1651	1565	310	3885	4027
200	6992	7090	311,131	2729	2876
201,021	4860	4904	204,024	373	431
112	18297	18131	222	28	81
003	345	317	303,033	358	495
202,022	5567	5403	312,132	4506	4335
103,013	2122	2151	400	831	801
210	314	373	105,015	2049	2133
211,121	13133	13164	401,041	1506	1549
113	2262	2415	214,124	2340	2435
300	590	692	223	2098	2144
212,122	8233	8068	402,042	286	258
203,023	9465	9568	115	1527	1471
301,031	5683	5822	313,133	1044	722
104,014	2831	2868	304,034	570	407
302,032	4090	3993	320	1139	1092

has to be included in the instrument function. In neutron diffraction with its in general Gaussian intensity distribution Cagliotti et al. (1958) have presented a formula widely used today also for X-ray experiments to describe the *FWHM* analytically (Eq. 3.10):

$$FWHM = (R\tan^2\Theta + S\tan\Theta + T)^{1/2} \tag{3.10}$$

This formula gives a reasonable description also in synchrotron diffraction experiments and Toraya has included this formula in his full pattern fitting program WPPD. The parameters R, S and T have to be determined with standard samples, normally silicon, quartz or lately CaB_4. Figure 3.11 depicts a plot of *FWHM* versus 2Θ for quartz.

3.2.8
Resolution of Profile Fitting

Better than any other method profile fitting can resolve reflections which are overlapped in the original experimental data. In general, two reflections can be resolved and their individual integrated intensities determined if two peaks are separated by more than halve the halfwidth. For lower symmetry crystals, or crystal with larger unit cell parameters it may be helpful to use constraints, for example to input the peak positions computed from known or previously determined lattice parameters, or using low angle better resolved reflections. Of course using a longer wavelength, say 1.2 Å, is a good choice in such cases. The analysis of an Ge-olivine crystal later in this monograph gives such an example. If the crystalline atomic parameters shall be determined in a crystal structure analysis the program POWLS allows one to lump together unresolved reflections in a peak cluster in the least squares routine FUNNY, if they cannot be separated sufficiently, or when the correlation between peaks calculated in the profile fitting program is too high.

The so-called five-finger triplet (212), (203), and (031) of quartz is generally taken as a measure to present the quality of resolution and of peak separation. This is demonstrated in Fig. 3.12. In ordinary, conventional laboratory powder diffraction using sealed X-ray tubes we have to deal with the α_1/α_2 doublet lines. Using monochromators with narrow line width and fine, e.g. long Soller slit systems the α_2 line can be sup-

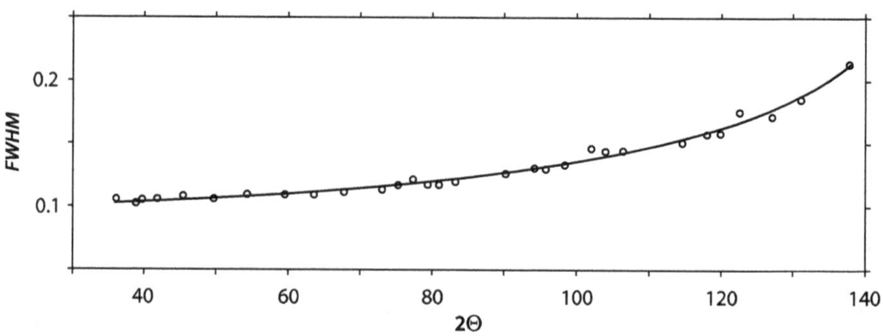

Fig. 3.11. Variation of *FWHM* versus 2Θ for quartz. The *circles* give the experimental *FWHM* values, the *solid line* represents the least squares fit with the parameters R, S and T

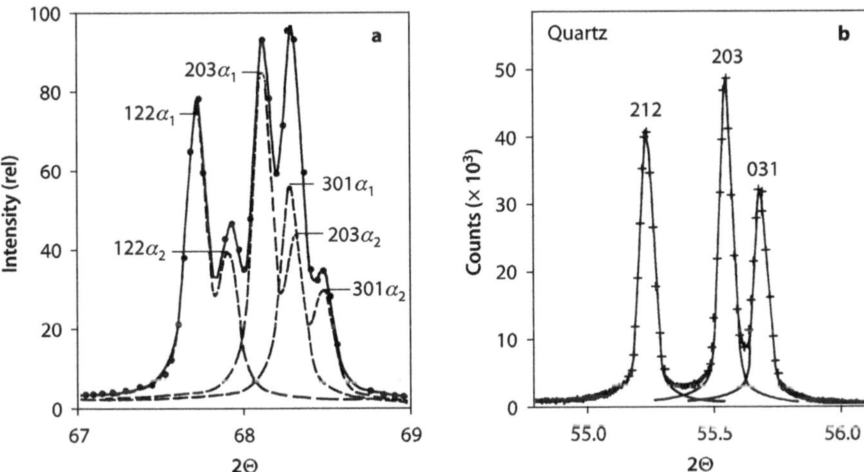

Fig. 3.12. Five-finger triplet (212), (203), and (031) of quartz; **a** measured with conventional Kα radiation and fitted with seven Lorentzian functions; **b** measured with synchrotron radiation

pressed to a large degree, however at considerable cost of intensity, and therefore experiment's time. There are mathematical procedures available, so-called line stripping programs. The best approach to record high resolution diffraction patterns is using monochromatic synchrotron radiation whenever it is available. Figure 3.12 shows the comparison between sealed tube radiation, Cu Kα, and synchrotron radiation. The conditions used here were what we can call "normal experimental conditions".

3.2.9
High Resolution Powder Diffraction – Lattice Parameters

The accurate determination of lattice parameters has aroused the interest of crystallographers for a long time. A comprehensive treatment has been published by Wilson (Wilson 1980). A serious attempt to compare the accuracy achieved in several laboratories in the world was sponsored by the International Union of Crystallography. Results have been published in 1960 (Parrish 1960). The main fact that emerged was that the agreement among laboratories was only 0.01%, which was about one order of magnitude lower than the precision claimed by the individual investigators and in the literature. The systematic errors were much larger than the random errors. Another open question was how to handle them in the various experimental methods. Two papers address these problems in detail (Parrish et al. 1987; Hart et al. 1990). In both cases experiments were made with synchrotron radiation on parallel beam powder diffraction stations, one at the SSRL, Stanford Synchrotron Radiation Laboratory, and the other at the Daresbury SRS synchrotron radiation source. Experiments with synchrotron radiation allow the use of strictly monochromatic radiation, a decisive advantage against laboratory measurements.

Many problems arise in the spectrum of the X-ray tube used in the laboratory. One major source is the angular separation of the K$\alpha_{1,2}$ doublet and its variation with 2Θ.

The profile asymmetry is another cause of error in the peak-angle measurements. One way out of most of these problems is in using synchrotron radiation parallel beam diffractometry. This technique will significantly increase the reliability of lattice parameter determination. Such a series of experiments have been performed by Parrish et al. (1987). They obtained patterns of very high resolution and used them to determine lattice parameters for a number of crystals (Table 3.5) to a very high degree of accuracy. The advantages of synchrotron radiation are obvious: Single wavelength, high intensity, wavelength selectivity and parallel beam technique. Parrish et al. used two vertical-scanning diffractometers, the first one for the silicon (111) channel monochromator, and the second one for the flat powder specimen. The specimen was continuously rotated around an axis normal to the surface to minimize crystallite-size statistical effects. One important experimental feature was a set of very long parallel slits: $l = 100$ and 365 mm. The full angular aperture of these slits was $\delta = 2\tan^{-1}(s/l)$, where s is the spacing and l the length of the thin foils. With $l = 100$ mm the result was $\delta = 0.17°$, and for $l = 365$ mm it was $\delta = 0.05°$. Figure 3.13 depicts the silicon powder pattern for 0.17° resolution with $\lambda = 1.0021$ Å. The lattice parameter was calculated to $a = 5.430825(20)$ Å.

The determination of peak positions is easier in synchrotron patterns, since the reflections are symmetrical. Nevertheless the mathematical description of the profiles is not necessarily standard. In this specific experimental setup for the 0.17° aperture the best function to describe the profiles was the sum of two Gaussian curves at the same peak angle. For the extreme long parallel slit system with 0.05° aperture a sum of a Gaussian and a Lorentzian curve gave the best fitting. Figure 3.14 gives typical reflections fitted with 75% Gaussian + 25% Lorentzian. As has been discussed before, the correct mathematical description of the profile is mandatory for reliable data. As has also been mentioned in synchrotron radiation experiments a double-Gaussian sometimes gives the best results.

In the investigation by Parrish et al. (1987) with synchrotron radiation peak positions were derived by profile fitting to an accuracy of 0.0001° in 2Θ. The average values of $\Delta d/d = \Delta a/a$ were directly calculated from the individual measurements. They ranged from 2×10^{-5} to 5.7×10^{-5}. For silicon a precision of 1 ppm was obtained with a standard deviation of the mean in the range 10^{-6}. The lattice parameters for quartz,

Table 3.5. Lattice parameters of some compounds determined by high precision measurements with synchrotron radiation ($T = 24.8$ °C)

Specimen	Lattice parameter (Å)	λ (Å)	Average $\lvert\Delta 2\Theta°\rvert$	Average $\Delta d/d$ ($\times 10^{-5}$)
Silicon	$a = 5.430825(20)$	1.002111	0.0020	2.3
Silicon and Tungsten	$a = 5.430825(46)$ $a = 3.165188(27)$	1.53467	0.0011	2.0
Quartz	$a = 4.912392(36)$ $c = 5.403848(70)$	1.28284	0.0021	3.9
Mg_2GeO_4	$a = 10.29171(55)$ $b = 6.02342(52)$ $c = 4.90592(20)$	1.7405	0.0018	5.7

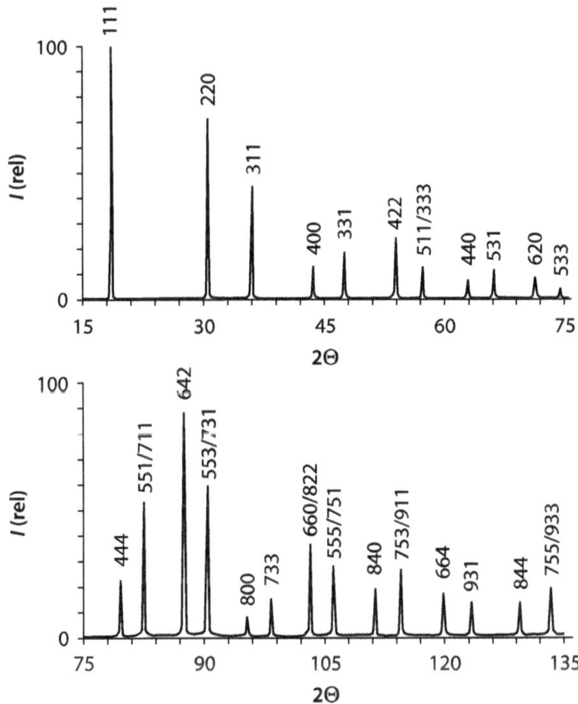

Fig. 3.13. Silicon powder diffraction pattern, 0.17° resolution, $\lambda = 1.0021$ Å

for example, are $a = 4.912392(36)$ and $c = 5.403848(70)$ Å calculated from 29 peaks. This means one part in 10 000, or about ±0.00001 Å. The average agreement between observed and calculated angles was $\Delta(2\Theta) = 0.0021°$, which equals a maximum precision of 0.002° of the peak position. The diffraction pattern for silicon, $a = 5.430825(20)$ Å, $\lambda = 1.0021$ Å, 0.17° resolution is depicted in Fig. 3.13.

In another set of experiments at the Cornell High Energy Synchrotron source CHESS Hastings et al. (1984) have reported results of extreme halfwidth in order to demonstrate possible applications of synchrotron X-ray radiation for high resolution powder diffractometry. They used a monochromatic beam from a perfect Si(220) double-crystal monochromator. Selected wavelength between 1.07 and 1.54 Å were taken for the experiments. An analyzer crystal behind the sample in the diffracted beam replaced the "receiving slit" and also eliminated fluorescence scattering etc. Analyzer crystals were Si, Ge, LiF and Al_2O_3. The goal was to investigate resolution and intensity characteristics. Using a Ge monochromator with $\lambda = 1.54$ Å he resolution $\Delta d/d$ at $2\Theta = 30°$ was 5×10^{-4}, falling to 2×10^{-4} at $2\Theta = 140°$ with the Si(111) analyzing crystal. Unfortunately this setup results in considerable peak asymmetry, which because of the intrinsically high resolution is much more evident at low angles than in the case of conventional diffractometers. With perfect Si crystals the resolution characteristics is governed mainly by the vertical divergence of the beam, roughly 0.1–0.2 mrad. The integrated reflectivity from a perfect crystal is rather low, since the intensity is proportional to $|F|$ rather than $|F|^2$ from a mosaic crystal. Also the step width required to record one diffraction peak must be extremely small, and consequently the time

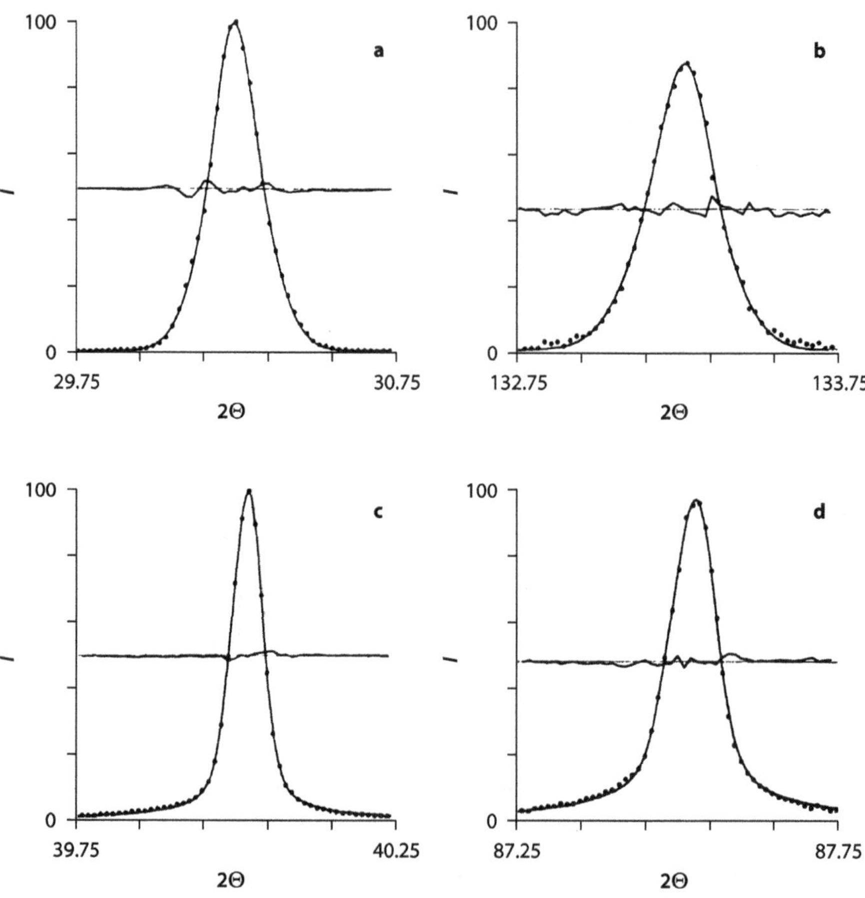

Fig. 3.14. Typical reflections of the silicon specimen, fitted with 75% G + 25% L. *Circles* are experimental points without smoothing Differences of experimental data and fitted profiles are shown at half height. The *FWHM* exceeding 0.17° in (**a**) and (**b**), and 0.05° in (**c**) and (**d**) is due to wavelength dispersion

needed to record a full pattern is rather long. In this case the step intervals were in multiples of 0.00125°, which required 800 seconds or about 14 minutes per degree, if we assume only one second per step. The halfwidth of the measured peaks was about 0.01° at $2\Theta = 20°$, compared to about 0.05–0.1° for a conventional instrument with Cu Kα radiation. Results of least squares fit of calculated and observed 2Θ values of 20 peaks for Al_2O_3 data taken with the Si(111) analyzer gave $\alpha = 4.7584(2)$ Å and $c = 12.9903(5)$ Å. The difference between observed and calculated angles $\Delta(2\Theta)$ was on average 0.002°.

Following the previous high precision determination of lattice parameters in silicon at the SSRL in Stanford Hart and co-workers performed a second study at Daresbury (Hart et al. 1990). This time they determined the lattice parameter of tungsten against silicon as an internal standard, yielding finally $a_W = 3.165188 (27)$ Å at 298 K. To avoid any possibility of drifts between runs the "internal standard", or "method of mixtures" technique was used (Klug and Alexander 1974). Tungsten is an ideal material, free from strain. The sample had a particle size of less than 5 μm. It was mixed with a sieved silicon fraction with a particle size 5–10 μm. The peaks were fitted individually with a Pseudo-Voigt function with excellent agreement. Repeated scans at the peaks were usually consistently fitted to 0.0001°

Conclusion

Both investigations (Parrish et al. 1987; Hart et al. 1990) show that lattice parameters for powder samples can be measured to a few parts per million using synchrotron radiation and parallel-beam X-ray optics. Unfortunately sufficiently accurate goniometers are rare. However the synchrotron laboratories offer stations which fulfill these requirements. Repeated scans (Hart and Parrish 1989; Cernik et al. 1990) show a reproducibility of 0.0001°, and given adequate counting statistics. The error curve is almost linear over the 2Θ range 10 to 80°, a range most users wish to use for structural studies and also for line broadening investigations.

3.3
Examples of Profile Fitting in the Two Stage Method

Some examples are given to demonstrate the wide range of applications of profile fitting with different programs in connection with the Two Stage method. Shown are examples of the application of PROFAN (Merz et al. 1990), of HFIT (Will and Höffner 1992; Höffner and Will 1991) and of FULFIT (Jansen E. et al. 1988). Shown in detail is also an extensive methodical investigation with quartz, which was done in order to check the power, but also the limits of the Two Stage method and POWLS, the program to refine crystal structures with the data from profile fitting.

Most programs used for the analysis of powder diffraction data were designed for main frame computers. Due to modern compilers and PC developments it is no problem nowadays to transfer established, large main frame refinement programs to personal computers. This is true for full pattern Rietveld routines and also for least squares programs based on integrated intensities like POWLS. However working with those programs on PCs is sometimes cumbersome because of lack of environment. This is for example the generation of starting a data file, or the inclusion of graphic displays of data files or patterns within the least squares programs. A publication by Jansen E. et al. (1994b) addresses these problems and describes a package of auxiliary programs. The programs are written in TurboPascal 6.0 and are aimed at IBM compatible PCs with a graphic adaptor like VGA. The wide potential of graphic manipulations is used to support the interactive operation at the PC.

3.3.1
Examples for PROFAN

PROFAN is an interactive program for on-line operation with a graphic display. It fits preselected profile functions, Gaussian, Lorentzian, Pseudo-Voigt, etc., into experimental peak clusters, or individual resolved peaks. The program operates *via menu* selections. It begins with a smoothing routine based on a fast Fourier transformation followed by background determination and subtraction. The profile fitting is performed stepwise and interactive on selected peak clusters after segmenting the pattern manually into suitable 2Θ sections. The expected peak positions are marked by the cursor, as indicated in Fig. 3.15 for the refinement of quartz. After each refinement step the operator can judge by the difference line (at the bottom of Fig. 3.15b) whether all peaks are properly accounted for, or additional peaks have to be added to the refinement. The operator can also change the profile function, if necessary. In heavily crowded patterns the *FWHM* parameter can be constraint. Figure 3.16 gives an example of a basically unresolved peak cluster of $TbAsO_4$ measured with neutrons at 0.4 K. After several steps and by adding additional peaks the cluster could be fitted with the six peaks indicated. The six peaks used to describe the experimental pattern are satellite peaks of a magnetic helical structure centered around the main peak forbidden in the nuclear structure and consequently with zero intensity. The final result is indicated by the difference plot at the bottom.

A third example is given in Fig. 3.17 of another neutron diffraction investigation of $Tb_{0.33}Y_{0.67}Ag$ taken at 16 K. In this example the area around the (111) region proved to be especially difficult to analyze because of considerable diffuse scattering under the peak. Expecting a helical moment configuration the analysis began with a Gaussian central peak with one satellite on each side, resulting in $R_{\mathrm{prof}} = 3.3\%$ (*a*). Continuing in (*b*) two satellites were placed on each side, with the *FWHM* still constrained. This

Fig. 3.15. Profile refinement of a section of a quartz pattern measured with synchrotron radiation (*data points*); **a** indicates the starting values by selecting peak maxima with the cursor, as marked by *crosses*. The starting profiles are also plotted; **b** depicts the resulting fitting with a *R*-value of 2.0%. The profile function has been a Pseudo-Voigt function

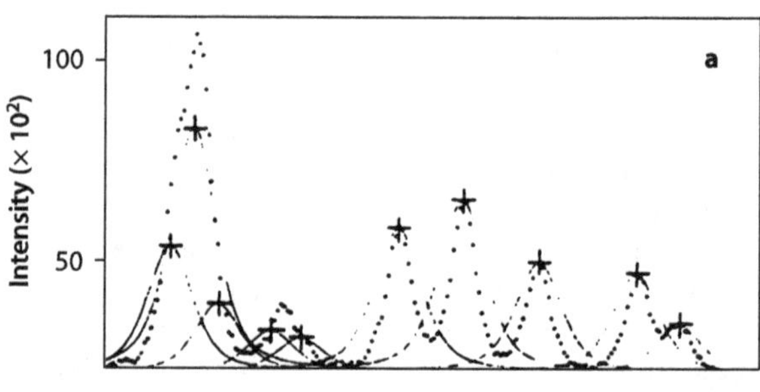

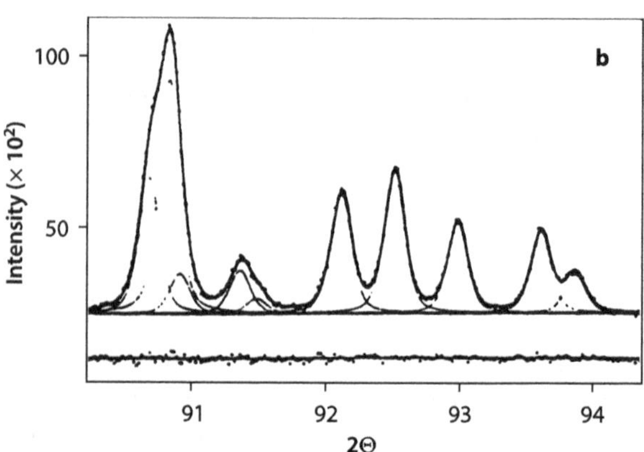

improved the picture to $R_{\mathrm{prof}} = 2.2\%$. All peaks in ($a$) and ($b$) are Gaussian. The diffuse contribution coming from diffuse Laue scattering was first described in (c) by a Lorentzian-shaped peak yielding $R_{\mathrm{prof}} = 2.0\%$. The final result is shown in (d) with $R_{\mathrm{prof}} = 1.6\%$. This is basically the sum of (b) + (c). This findings show a gradual transition in the structure from long-range magnetic order resulting in satellite reflections to a spin-glass behavior with decreasing x in $\mathrm{Tb}_x\mathrm{Y}_{1-x}\mathrm{Ag}$, as revealed in diffraction patterns with different values of x. Diffraction patterns coming from such compounds

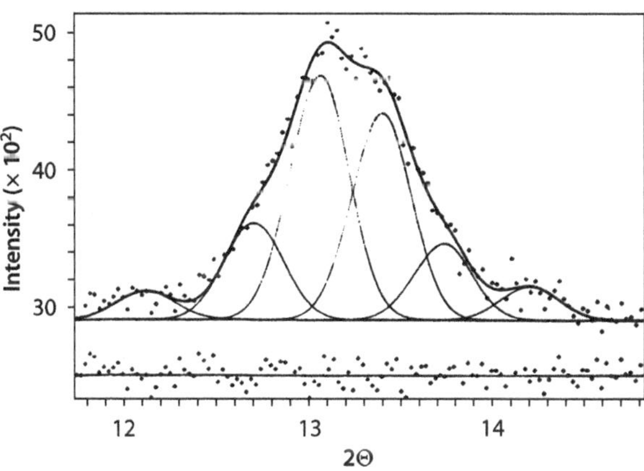

Fig. 3.16. Profile refinement of a peak cluster of a section of TbAsO_4 measured with neutrons at 0.4 K. The *FWHM* was constraint after calibration with a standard sample. The best fit ended with $R_{\mathrm{prof}} = 2.3\%$

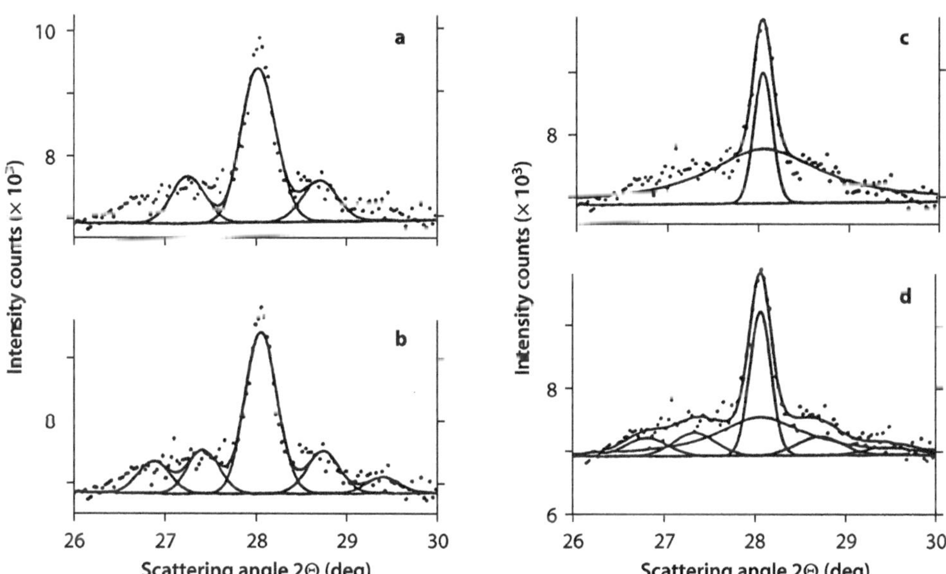

Fig. 3.17. Profile analysis of a region around (111) of $\mathrm{Tb}_{0.33}\mathrm{Y}_{0.67}\mathrm{Ag}$

can reliably only be analyzed with interactive profile fitting, especially if Gaussian and Lorentzian shaped contributions are overlapping each other.

3.3.2
Examples for HFIT

HFIT has been developed for the analysis of in general terms "difficult" diffraction patterns coming from experiments, for example with synchrotron radiation with the specimen under pressure. Figure 3.18 shows the bloc diagram of HFIT: After input of the data the program performs on request smoothing of the pattern *via* a fast Fourier algorithm (Press et al. 1988). It then subtracts the background using orthogonal weighted polynomials represented as Tschebyscheff-like polynomials. In deviation to normal Tschebyscheff polynomials here each data point is given its own weight in order to separate background and signal point. In abnormal cases, where the background changes abruptly when going through an absorption edge, as shown in Fig. 3.19, Akima polynomials are used (Akima 1970). As the next step the program asks the on-line operator to divide the pattern into sections, e.g. zoning the diagram, similar to the procedure in PROFAN. The sections are then searched each for peaks and fitted with a preselected profile function. The profile refinement is based on the Levenberg-Marquardt nonlinear least squares algorithm.

Fig. 3.18. Block diagram of HFIT and logical flow scheme

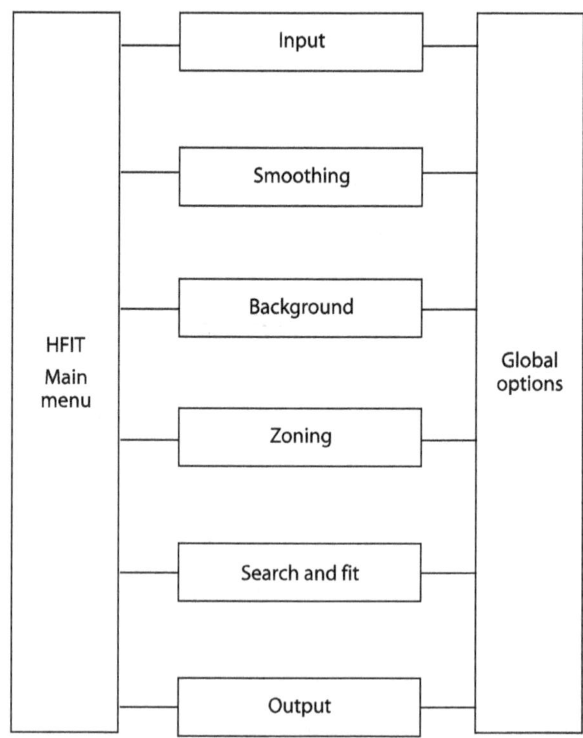

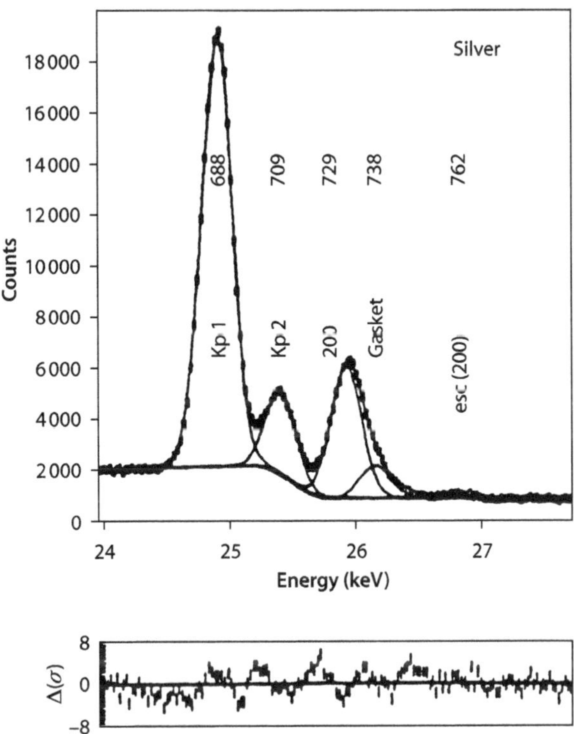

Fig. 3.19. Section of the diffraction pattern of silver measured with energy dispersive synchrotron diffraction going through the absorption edge of silver with an abrupt change of background

3.4
POWLS

3.4.1
Structure Refinement with POWLS

At this stage when using the Two Stage method we have integrated intensities of all reflections of the diffraction pattern derived from the profile fitting procedure. In addition we assume that the reflections are indexed. Those data are used as input values for the second step in the Two Stage method, for example for structure refinements. For refinements of (basically known) crystal structures the very general program POWLS (Will 1979; Will et al. 1983b) is used. Of course any other least squares program may be adapted. POWLS has been used over more than 30 years with great success in X-ray, synchrotron and neutron diffraction work.

3.4.2
The Program POWLS

POWLS (POWder Least Squares) is a re-iterative nonlinear least squares program based on a first order Taylor expansion formalism (Hamilton 1964a). It is written in FOR-

TRAN and has been adapted for use on personal computers. A block diagram of the program is shown in Fig. 3.20. Quantities designed as "observations" may be structure factors $F(hkl)$, integrated intensities $I(hkl)$, or any function of intensity, for example in neutron diffraction. This makes POWLS for example especially useful in neutron diffraction when dealing with magnetic structures.

POWLS is characterized by three main sections and the output routine:

- MODEL   Definition and preparation of the structural model
- OBSERVATION   Setup of the observations with its standard deviations and correlations, if available
- REFINEMENT   Performing the least squares refinement in several cycles
- OUTPUT   Output

The refinement is carried out on a general observed quantity G(obs), a vector or a one-column matrix with n components g_i. The values G, actually G^2, are defined by the operator:

Least squares calculation: $\Sigma w\,(G^2(\text{obs}) - G^2(\text{calc}))^2 = \text{Minimum}$ (3.11)

with (for example):

a $G^2 = I$ (Intensity)
b $G^2 = \Sigma j\,F^2$
c $G^2 = \Sigma <q^2> p^2 j F^2$ (for magnetic structures)

G can be any quantity. In common applications G are integrated intensities $I(hkl)$ of individual reflections, or groups of overlapping reflections, derived in the first step of the profile fitting procedure. The handling of overlapping peaks or groups of peaks are the main advantage of the program and this makes POWLS so powerful. In many structures, e.g. space groups we are confronted with intrinsically overlapping and unresolved reflection with individual intensities, like (333) and (511) in cubic symmetry, or (10.1) and (01.1) in quartz and similar structures. In other cases there are often occasions where we wish to treat several reflections as a group, for example when we have strong or unacceptable correlations in the first step of the profile fitting program, or if the experimental conditions are such that the individual peaks are not resolved. This may occur when we have poor data or poor counting statistics, for example due to sample size or absorption, or when the step width in the experiment is too coarse for a meaningful profile analysis. In such cases the program POWLS allows one to treat the overlapping region either as one observation or alternatively as a set of individual observations restricted by correlations, known from the first step of the analysis, the profile fitting.

Sometimes it is preferred to use the reduced value $\Sigma j F^2$ as "observations" in the least squares calculation, e.g. the intensities divided by the Lorentz and, if applicable, polarization factors, for example in refinements magnetic structures from neutron diffraction experiments. Specific cases of neutron diffraction are treated later in this monograph. Any type of parameter can be used. In addition to the usual positional and thermal parameters, occupancy factors, magnetic parameters (for example magnetic moment value p and the moment orientation $<q>$ in neutron diffraction), form

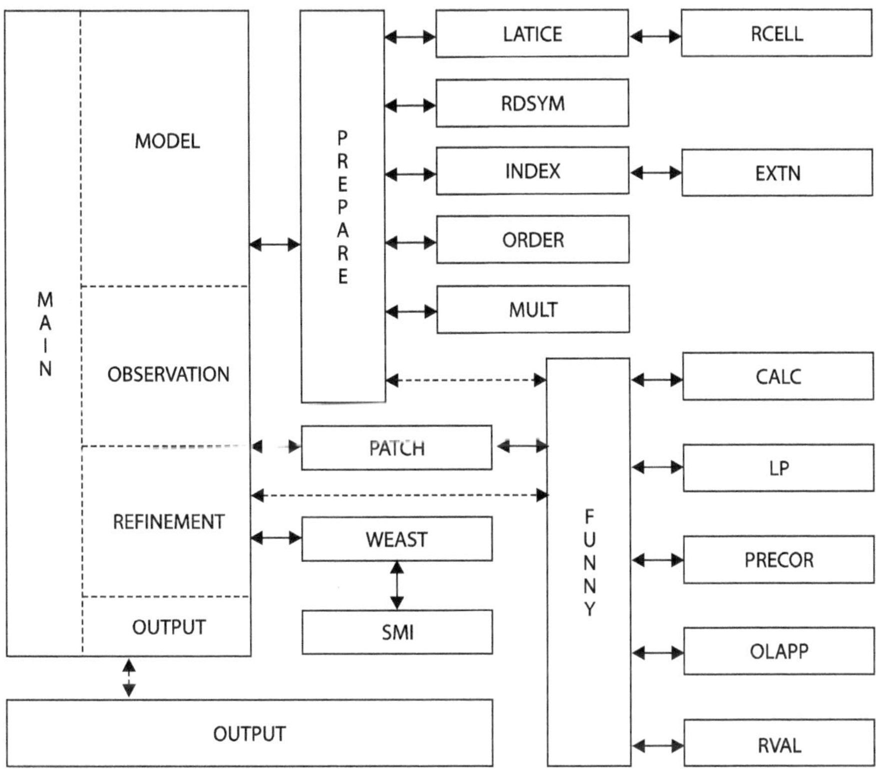

Fig. 3.20. Block diagram of POWLS

factors, unusual descriptions of thermal motions, correction terms for preferred orientation, etc., are common and are incorporated in POWLS.

The refinement is carried out on a general observation quantity G(obs) with the components g_i, which is approximated by a vector $C = G(\text{calc})$ or a one column matrix with n components approximated by a similar vector $C = G(\text{calc})$ with the components c_i calculated from a set of parameters described by a parameter vector P with $m < n$ components p_j. The quantity to be minimized is then in general terms

$$\mathbf{M} = \mathbf{R}^{\mathrm{T}}\mathbf{W}\mathbf{R} \tag{3.12}$$

with $\mathbf{R}$ the vector of the residuals, with the components

$$r_i = g_i - c_i(\mathbf{P}) \tag{3.13}$$

$\mathbf{R}^{\mathrm{T}}$ is the transposed matrix of $\mathbf{R}$, and $\mathbf{W}$ is an (estimated) weight matrix similar to that used in single crystal least squares procedures. $\mathbf{W}$ is proportional to the inverted variance-covariance matrix of $\mathbf{G}$.

POWLS reduces the general nonlinear least squares problem to a linear form by using a Taylor series expansion. We assume, that the original parameter vector $\mathbf{P}$ can be replaced by a better parameter vector $\mathbf{P'}$ obeying the equation

$$g_i = c_i(\mathbf{P'}) + r_i' \tag{3.14}$$

with r_i' the components of a remaining residual vector $\mathbf{R'} < \mathbf{R}$. The mathematical problem is to find the adjustment vector $\mathbf{X}$ given by

$$\mathbf{P'} = \mathbf{P} + \mathbf{X} \tag{3.15}$$

The first order Taylor expansion of c_i about the point $\mathbf{P}$ gives

$$c_i(\mathbf{P'}) = c_i(\mathbf{P}) + \Sigma j\,(\partial c_i/\partial p_j)x_j \tag{3.16}$$

Defining a matrix $\mathbf{A}$ with the elements

$$a_{ij} = \partial c_i / \partial p_j \tag{3.17}$$

and referring to Eqs. 3.13 and 3.14 we obtain

$$\mathbf{R} = \mathbf{A}\mathbf{X} + \mathbf{R'} \tag{3.18}$$

This is an overdetermined system of linear equations, which can be handled by a least squares formalism meeting the requirements of Eq. 3.11. This procedure is described by Hamilton (1964), and leads to

$$(\mathbf{A}^{\mathrm{T}}\mathbf{W}\mathbf{A})\mathbf{X} = \mathbf{A}^{\mathrm{T}}\mathbf{W}\mathbf{R} \tag{3.19}$$

This is the matrix equation formed and solved by POWLS. The Taylor series expansion of Eq. 3.16 is an approximation and is valid only for small x_j. In general, several refinement cycles, normally only a few, 3 to 5, are necessary in order to find the best $\mathbf{P}$.

In the cases treated here the g_i are identical with the integrated intensities $I(hkl)$, or sums of I in a group of unresolved reflections, or in some cases $\Sigma j F^2$. The g_i are derived from the profile fitting procedure in Step 1. They are not correlated, and therefore $\mathbf{W}$ is diagonal.

The quantity to be minimized is then in crystallographic terminology

$$\sum_i w_i \left| g_i(\mathrm{obs}) - (1/K)\cdot c_i(\mathrm{calc}) \right|^2 \tag{3.20}$$

where K is a scale factor. The proper weighting system used in POWLS is the inverse variance-covariance matrix calculated in the course of the profile analysis. When the g_i are not correlated, $\mathbf{W}$ is diagonal. The degree of correlation depends very much on the profile analysis, and thereby especially on the intervals between the data points. High correlations are encountered consequently when peaks are too close together. *POWLS provides the means to avoid such problems by treating overlapping reflections in the subroutine OLAPP.*

In most cases in practice the weight matrix $\mathbf{W}$ is not available. In those cases we assume the components w_i to be proportional to the reciprocal variances $1/\sigma^2$ of the g_i. Because of the unknown contribution from the background it is appropriate to add a constant value, yielding

$$\sigma(hkl) = \text{Constant} + \sqrt{I(hkl)} \tag{3.21}$$

The original version in 1963 was written for main frame IBM computers. Meanwhile the program has been modified and especially it has been adapted to PCs. Refinement of a data set of quartz for example now takes about 5 seconds. For a FORTRAN listing of the program see Will et al. (1983a,b). There is information of the data input, information on the parameter refinement switches and an example of a practical refinement of α-quartz.

The program MAIN first reads the input data and organizes the program with respect to the subroutines. A subroutine XTAL organizes the calculation of the structure factors $F(hkl)$ for all reflections allowed by the space group. Form factors are stored in the computer and are calculated from tables taken from the *International Tables for X-Ray Crystallography*. The crystallographic extinction conditions by symmetry are calculated from space group conditions and are also permanently stored. The space group number, together with a set of atomic coordinates is thus sufficient as input.

The structure factors are calculated in the problem oriented subroutine FUNNY, where also the vectors of the reduced intensities G are determined. FUNNY is the central subroutine of the whole program POWLS. It has a loop through all reflections allowed by the space group conditions and a second loop through all the atoms per unit cell. FUNNY also provides the user with the possibility to calculate (and refine), if necessary, parameters not commonly used, like partial site occupancies or magnetic properties in neutron diffraction. It then calls PRECOR, which is designed to make corrections for preferred orientation, i.e. "non-random" particle distribution in the specimen (see Sect. 3.5), and OLAPP to calculate theoretical intensities with the choice to lump together overlapping reflections, may it be intrinsically like (333)/(511) in cubic or (10.1)/(01.1) in crystals with rhombohedral symmetry, or in cases of a small group of very closely spaced reflections to be separated by profile fitting only with large correlations.

WEAST is a weighted least squares routine and SMI a symmetric matrix inversion routine. This finally goes to OUTPUT to put the refined data in an output file and/or print the results.

For the refinement residuals a R-value is calculated. The correct structural refinement residual is the so-called Bragg R-value close to the residual R used in single crystal diffractometry. Only the Bragg R-value gives a reasonable judgement of the refinement. For a detailed discussion of R-values we refer to Sect. 3.6.

3.4.3
Preparation of a Model Structure

The preparation of the model structure is achieved through the subroutine PREPAR controlling the following five subroutines LATICE with RCELL, RDSYM, INDEX with EXTN, ORDER and MULT (see block diagram, Fig. 3.20).

- LATICE reads the cell parameters, in direct or reciprocal values, and the wavelength. For X-ray data with the symbols of the X-ray tube, like Cu, Mo etc. are sufficient. It calls RCELL for calculating reciprocal cell parameters, if needed
- RDSYM (for read symmetry) reads a space group number for the space group given by the operator and steers the access to the data file containing the symmetry information taken from the *International Tables for X-Ray Crystallography*. RDSYM calculates a complete set of symmetry positions including centrosymmetric values for possible further application in magnetic structure analysis for neutron diffraction work
- INDEX generates the Miller indices *hkl* for the corresponding Laue group (taken from RDSYM). It saves those which are not affected by extinction rules defined in EXTN
- ORDER orders the Miller indices according to decreasing *d*-values
- MULT calculates the appropriate multiplicity values for each index triplet (*hkl*)
- PREPAR performs the calculation of trigonometric tables and reads the parameters *P* necessary to describe the model: They include the scattering lengths for neutrons or form factor tables for X-rays, which are already stored permanently in a separate data file. This form factor file can be ignored and substituted by individual form factor tables. Next are starting parameters for the atomic positions, temperature factors and several overall parameters like the scaling factor or a parameter for calculating preferred orientation
- PATCH enables the operator to include constraints between parameters for example with symmetry related positions or temperature coefficients. This is similar to the Busing-Levy original least squares programs used for single crystal structure analysis. With these informations a list is generated and printed containing *hkl*, 2Θ positions, *d*-values and model intensities

The intensity calculation of $C(P)$ itself is done in the subroutine FUNNY with several additional and individual subroutines defined by the operator according to the problem to be solved. The schematic flow diagram of FUNNY is shown in Fig. 3.21. First an outer loop runs over the involved reflections described by an index vector $\mathbf{h}$ (= *hkl*). The structure factor $F(\mathbf{h})$ is calculated as

$$F(\mathbf{h}) = \sum_j f_j \exp(2\pi i\, \mathbf{h}^T \mathbf{x}_j)\exp(-\mathbf{h}^T \mathbf{B}_j \mathbf{h}) \tag{3.22}$$

With the index j summing up all atoms of the unit cell, f_j the form factor or neutron scattering length, $\mathbf{x}_j$ the positional vector and $\mathbf{B}_j$ the matrix describing the anisotropic vibration. For convenience the calculation of $F(\mathbf{h})$ is split up into two loops:

$$F(\mathbf{h}) = \sum_p f_p(\mathbf{h})\sum_s \exp(2\pi i\, \mathbf{h}^T \mathbf{x}_{p,s} - \mathbf{h}^T \mathbf{B}_{p,s}\mathbf{h}) \tag{3.23}$$

Loop 1　　Loop 2

The first summation p is over the Wyckoff positions p. It is performed in the subroutine FUNNY and includes the summation over the corresponding symmetry-equivalent positions s in the subroutine CALC. The scattering intensities are calculated as

Fig. 3.21. Schema of the subroutine FUNNY

Subroutine FUNNY

Set up constants

Define variables $(x, y, z, B, K, \ldots)$

DO HKL

   DO j for ATOMS

      Calculate formfactors $f(j)$

      Calculate $F(hkl)$

$$F(hkl) = f(j) \cdot w(j) \cdot A(j) \cdot \exp(B(j) \cdot T^2)$$

   END j

END HKL

CALL PRECOR   (subroutine for preferred orientation)

CAL OLAPP   (subroutine for overlapping peaks)

RETURN

$f(j)$ = form factor

$w(j)$ = occupancy number (according to Wykoff)

$B(j)$ = temperature factor, isotropic or anisotropic

T = $\sin \Theta / \lambda$

A = geometrical structure factor listed in and taken from the International Tables of Crystallography

$$I(\mathbf{h}) = Sc\, Mu(\mathbf{h})\, Lp(\mathbf{h})\, Po(\mathbf{h})\, |F(\mathbf{h})|^2 \tag{3.24}$$

with:

- Sc: defining a scale factor
- Mu: the multiplicity of the reflections (taken automatically from the international tables stored in the program)
- Lp: the Lorentz-polarisation correction depending on the diffraction geometry and stored in the program (it can be selected by setting the appropriate switch in the input file)
- Po: the preferred orientation correction performed by the function subroutine PRECOR

3.4.4
Observations

This second part of POWLS is activated when the program is used for structure refinement from the observed intensities. Thereby POWLS is independent of the method by which these intensities have been obtained: By profile analysis and profile fitting, by individual integration or just by estimation, taken for example from the peak heights.

This is a major advantage of the Two Step method compared with the Rietveld method. Even taking values from the JSPD powder files gives in general good results, and it may indicate besides other factors the quality of the data in the file. The observed values are processed using information from the user about experimental error margins and about correlations coming mainly from overlapping and unresolved reflections. All reflections *hkl* of the structure allowed by symmetry and space group are calculated by the program, independent of observed or not observed and printed in a first run. They are individually numbered in the output of PREPAR. Non-observed reflections are in structure refinements a sensitive indications about the correctness of the structure, since non-observed reflections must be calculated at least as very weak in the refinement. They are now attached in the input file by the user to the observations by assigning the (*hkl*)-number(s) to the corresponding observation. Error margins may be assigned to each observation. When correlations between reflections are available, for example from a profile fitting program, they should be used as an input. Error margins and correlations are processed to calculate a non-diagonal weight matrix $\mathbf{W}$. In a final list the observation are printed together with the attached *hkl*-numbers.

3.4.5
Refinement

With the informations about a structural model and the observations available the actual parameter refinement begins. N cycles, defined at the start, are processed. There is a parameter switch for each cycle indicated in the "switch card", so the parameters can refined stepwise individually. As a rule, the first two cycles should be limited to refine just the scale factor, which by this procedure can be set in the INPOUT file as a coarse value. The number of activated switches determines the dimension of the actual refinement calculation.

In order to calculate the intensities c_i of the model for forming the matrix $\mathbf{C}$ the MAIN program activates the subroutine FUNNY to obtain individual model functions with respect to the parameter matrix $\mathbf{P}$. The residual matrix $\mathbf{R} = \mathbf{G} - \mathbf{C}$ containing the deviations between the matrix $\mathbf{G}$ of the observations and the model matrix $\mathbf{C}$ is stored for further use in the next cycle.

The calculation of the partial derivatives is performed by changing the parameters to be varied one by one by small increments ∂p_i and forming a neighbour matrix $\mathbf{C}'$. The partial derivatives c_i defined by $(c_i'-c_i)/\partial p_j$ are stored in the matrix $\mathbf{A}$.

Knowing $\mathbf{A}$, $\mathbf{R}$ and the weight matrix $\mathbf{W}$ the subroutine WEAST is activated to calculate the matrix $\mathbf{N} = \mathbf{A}^{\mathrm{T}}\mathbf{W}\mathbf{A}$ of the normal equations and the vector $\mathbf{Y} = \mathbf{A}^{\mathrm{T}}\mathbf{W}\mathbf{R}$. For solving the equation $\mathbf{N}\mathbf{X} = \mathbf{Y}$ the subroutine SMI performs a matrix inversion $\mathbf{N}$ to $\mathbf{N}^{-1}$. The matrix of the parameter adjustments is obtained by $\mathbf{X} = \mathbf{N}^{-1}\mathbf{Y}$.

3.4.6
Preferred Orientation

Powder diffraction is based on a statistical distribution of many crystallites. To determine the true relative intensities a completely random orientation of the crystallites is required by theory. Specimen with completely random orientation however are hard, if not impossible to prepare. Therefore corrections are necessary. After taking precautions and care in the preparation of the specimen and in the experiments a mathematical correction is necessary in the course of the refinement and such corrections are provided in all refinement programs in use. POWLS allows to make corrections in a subroutine called PRECOR, where a quantity GP is added to the list of variables. POWLS has three formulas available listed in Sect. 3.5.3.

On the basis of a modified version of POWLS Valvoda et al. (1996) have tackled the problem of correcting for preferred orientation of crystallites by using a simple empirical function. This treatment is useful and has been applied for strongly textured materials, especially with fiber texture in the specimen. They call their method "joint texture refinement". The method is based on simultaneous fitting of calculated intensities to the whole set of intensities measured at several inclinations χ of the specimen. The method has been applied to brucite, $Mg(OH)_2$, and to a thin film specimen of zinc.

Zinc had a strong (001) fiber texture. Taking measurements at the ordinary Bragg-Brentano geometry, $\chi = 0$, only two reflections 002 and 004 were visible. Two further measurements were taken at $\chi = 35$ and 46°, which are the angles of the (103) and (102) planes with respect to the (001) plane and these measurements enhanced the intensities from these two planes. The texture could then be determined with a modified program POWLS yielding the texture parameters and good agreement between I(obs) and I(calc). The R-Bragg values were between 0.53 and 1.38% for the three measurements.

3.5
Specimen and Preferred Orientation

3.5.1
Specimen Preparation

It is evident that the preparation of the specimen is crucial for achieving good results. It is a difficult art and it requires a great deal of patience to prepare specimens of suitable quality. Time spent on specimen preparation is well spent. A failure in the specimen can not be compensated by computational corrections afterwards.

Ideally there should be a sufficient number of randomly oriented crystallites in the specimen to provide correct intensities for all reflections. The effective number having the correct orientations is determined by

- particle size,
- multiplicity factor, and
- instrument geometry.

Only about 0.1 cm^3 volume takes part in X-ray powder diffraction. Because in the Bragg-Brentano geometry, which is normally used, the surface layer is the main contributor of the diffraction, the final surface finish is an important factor. Also inter-particle micro-absorption can exert a major effect on the results. Particle size is above all the crucial parameter. Small particles in the range of 5 to 10 μm are required for good results. Even smaller particles of 1 μm are still better, if they can be obtained. It is realized that it is difficult to extract particles of that fraction from a sample, but it can be obtained. For the results shown in Figs. 3.34 and 3.36 the silicon powders were sifted through 5, 10, 20 and 30 μm micromesh using an acoustic sifter to provide the shown size fractions. It can be done for example by intensive grinding and sedimentation in water in the case of quartz (see Will et al. 1983a). A technique used by Parrish et al. (Will et al. 1983a) in an investigation of quartz fine grained specimens were obtained by intensive grinding and sedimentation in water.

3.5.2
Random Distribution

Powder diffraction is based on a statistical distribution of many crystallites. To determine the true relative intensities a completely random orientation of the crystallites is required. Such specimens however are hard, if not impossible to prepare. Therefore precautions are necessary in the experiments and corrections through calculations must be made in the evaluation. This difficulty is not restricted to powders with good cleavage, like mica or calcite. In any specimen there will be some deviation from a perfectly random particle distribution. This we prefer to call "*non-random distribution*". It will always be present, even in samples such as silicon or quartz.

For the ensuing discussion we must distinguish three border line cases:

- Real texture in the sample, for example in rolled metal sheets or in rocks. Here the crystallites show a distribution depending on the treatment of the sample, for example by rolling, or on conditions during crystallization or metamorphism in rocks. This is a special line of science named "texture analysis". Example are presented in Sect. 3.9
- Preferred orientation in its true sense is in general the result of cleavage properties in the crystals, which is found for example in some minerals like mica, calcite, dolomite, feldspar etc.
- *Non-random distribution* because it is impossible to achieve 100% randomness in the specimen

The goal of powder diffraction analysis is the determination of the diffracted intensities as accurate as possible. Since in reality we have deficiencies in the specimen, a number of steps have to be followed:

- *Specimen preparation* must be very careful with small uniformly distributed particles around 5 μm. Spherical samples would be the best approach to a random distribution. Several techniques are proposed for creating "nearly random" samples, like the one proposed by Snyder (personal communication in 1990). But his prescriptions are difficult to follow on a routine basis

- *Experiments:* Rotation of the specimen, be it flat as in Bragg-Brentano geometry, or cylindrical as in Debye-Scherrer technique, is actually mandatory in order to reduce preferred orientation at least to some degree. This will improve the problem in one dimension, but not solve it
- *Corrections* in the data evaluation or in structure refinement

3.5.3
Calculational Corrections

Corrections during the refinement have already been proposed and provided by Rietveld in his first program version. Today every least squares structure refinement program has such provisions. The program POWLS has provisions for corrections in a subroutine called PRECOR, where a quantity GP is added to the list of variables. During the refinement intensities are calculated as

$$I(\mathrm{corr}) = I(\mathrm{obs})\,P(\Phi) \tag{3.25}$$

$P(\Phi)$ is a parameter to correct the observed intensities for "preferred orientation" effects. For such corrections three formulas are offered:

- Gaussian (Rietveld 1969)

$$P(hkl,\Phi) = \exp(-GP\,\Phi^2) \tag{3.26}$$

- Exponential (Will et al. 1983)

$$P(hkl,\Phi) = \exp(GP(\pi/2 - \Phi^2)) \tag{3.27}$$

- Dollase (1986)

$$P(hkl,\Phi) = (GP^2\cos^2(\Phi) + \sin^2(\Phi) / GP)^{-3/2} \tag{3.28}$$

GP is the correction parameter, which is a variable in POWLS, Φ is the acute angle between the scattering vector, defined by (hkl), and a vector (HKL) defined by the operator as the "preferred orientation" vector. In X-ray diffraction with flat samples Eq. 3.27 will be valid, where we assume crystallites with their axis vertical to the scattering vector (Bragg-Brentano geometry). Equation 3.26 assumes crystallites in cylindrical sample holders (neutrons).

Equation 3.28 was proposed by Dollase (1986). It gives the best results, if considerable corrections are required, for example with specimens containing for example calcite or other crystals with pronounced cleavage properties. The formalism by Dollase is based on a paper by March (1932), who already in 1932 considered this problem and derived a formula for correction. Equation 3.28 leads to good results when sizeable preferred orientation is encountered, even unintentionally. See for example Will et al. (1988b).

The correction depends naturally on the scattering angle 2Θ, but also on the orientation of the reflecting plane (hkl) with respect to a plane (HKL) defined as the "preferred orientation" plane. Except in crystals with known cleavage planes, this plane (HKL) is not known beforehand and must be found by "trial and error" until the lowest R-value is obtained. Will and Parrish (Will et al. 1983a) studied this phenomena with silicon and quartz. Table 3.6 gives results for the case of a silicon specimen. Here all possible planes were checked, including planes not allowed by symmetry, with the results shown. (HKL) = (100) gave finally the best value with $R = 0.86\%$ versus $R = 3.5\%$ without correction.

That the term "preferred orientation" is misleading, or is even wrong, can be seen in the specific case of two silicon specimens, which were prepared from a sample provided by the National Bureau of Standards as Reference Material 640. Both specimens used fractions of the same sample which passed through 10 μm micromesh (Parrish and Huang 1983). The two specimens were independently prepared and measured and refined. In the first case, listed in Table 3.7, the R-value could be brought down from $R = 4.4\%$ to $R = 0.74\%$ by correcting the intensity data by a quantity GP of Eq. 3.27. This was surprising first because no cleavage plane exists in silicon, second and more over the "preferred orientation plane" was found as (111). And thirdly in a second specimen from the same sample the "preferred plane" HKL was (100) with $R = 0.86\%$ versus 3.5% before correction (Will et al. 1983a). There is no scientific basis behind these planes found. The name "*non-random distribution*" is appropriate. The structural parameters are not affected by the correction.

3.5.4
Crystallite Size Effects

Particle size is a further fundamental and critical factor affecting absolute and relative intensities contributing more than any other factor to the effect of "*non-random distribution*". This is especially critical in Bragg-Brentano geometry, which is based on a divergent beam impinging onto the flat specimen, and then being focussed onto the detector. Only about 0.1 cm^3 volume takes part in X-ray powder diffraction, and in the Bragg-Brentano geometry the surface layer is the main contributor. The problems arise from large particle sizes and micro-absorption. It requires therefore a great deal of patience to prepare samples of suitable quality. As discussed rotation of the specimen is a first and mandatory provision to obtain reliable intensity data. But second the size and size distribution of the particles in the specimen is of similar importance. It will become even worse when using synchrotron radiation because of its high collimation.

To demonstrate this effect specimens with different particles sizes were studied, first in the laboratory with Cu Kα X-rays. Figure 3.22 shows the variation of peak intensities of the (10.0) peak for four quartz specimens when rotated around the azimuth orientation ϕ at a reduced speed of 1/7 r min^{-1}. The peak intensities vary drastically in the way as large particles not removed by sifting come into reflecting position and lead-

Table 3.6. Result of the refinement of silicon by including a correction for "preferred orientation" in POWLS. (*HKL*) is the plane treated as the "preferred orientation" plane

hkl	*R* (%)	*hkl*	*R* (%)	*hkl*	*R* (%)
No correction	3.50	100	0.86	531	2.44
111	1.86	331	1.73	442	1.69
220	2.02	422	2.43	620	1.25
311	2.01	511	1.36	533	2.40

Table 3.7. Intensities *I*(obs) and *I*(calc) without and with correction for "preferred orientation" in the silicon sample S12. (111) is taken as the direction (*HKL*) for the plane treated as the "preferred orientation" plane

hkl	*I*(obs)	*I*(calc) no correction	Ratio	*I*(calc) with correction	Ratio
111	7729	7569	1.02	7729	1.00
220	4639	4932	0.94	4648	0.993
311	2829	2897	0.97	2828	1.00
400	758	759	0.998	726	1.04
331	1242	1127	1.10	1187	1.04
422	1682	1549	1.08	1702	0.99
511/333	940	879	1.07	965	0.97
440	628	559	1.12	620	1.01
R-value (%)		4.4		0.74	

ing to high spikes. In the usual rapid rotation of about 60 r min^{-1} this effect is smeared out but nevertheless present.

Figure 3.23 depicts a series of similar recordings of intensity variations of silicon specimens from samples with different grain sizes rotated fast and slowly around the specimen normal. With fast rotations as in common experiments (at the right side of the patterns) this effect is smeared out, but naturally is still present giving incorrect intensities despite rotation. Rotation averages the in-plane preferred orientation, although it has virtually no effect on the preferred orientation parallel to the specimen surface. Mathematical corrections are needed to reduce this deviations. This effect is even worse when using synchrotron radiation because of the high collimation and this is demonstrated in three representative recordings in Fig. 3.24. With particles of 5 μm or smaller there is basically no effect. Larger particles on the other hand can give extremely high spikes.

Particle size

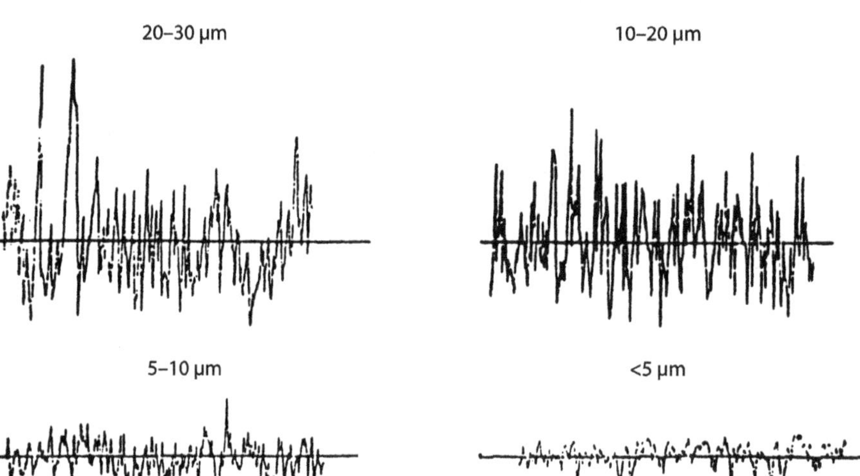

Fig. 3.22. Effect of particle size on the intensity of the quartz peak (10.0) during slow azimuth rotation ϕ, measured with Cu Kα X-rays

3.6
R-Values

3.6.1
General Remarks

To judge the quality of analysis and structural calculations one requires a generally accepted quantity. In crystallography this is the *R*-value, which we may call a residual index (less desirably a reliability index) (Stout and Jensen 1970):

$$R_w = \frac{\sum |\Delta F|}{\sum |F_{obs}|} = \frac{\sum \left\| w F_{obs} | - | F_{calc} \right\|}{\sum |w F_{obs}|} \qquad (3.29)$$

In single crystal structure analysis this quantity is defined and calculated as the differences between calculated and observed structure factors $F(hkl)$. Unweighted and weighted values are used, R and R_w, with w the weight $w = 1 / \sigma^2$ and σ the e.s.d (estimated standard deviation) to consider experimental uncertainties.

Equation 3.29, as used in the refinement of single crystal data, is a straight forward calculation without ambiguities. With the advent and the continuing interest in crystal structure refinement from powder diffraction data and later also in crystal structure analysis of unknown structures researchers began to present their results with the same, or at least similar reference numbers. So we find R-values in all publications, unfortunately however with very different and quite arbitrary definitions. It must be

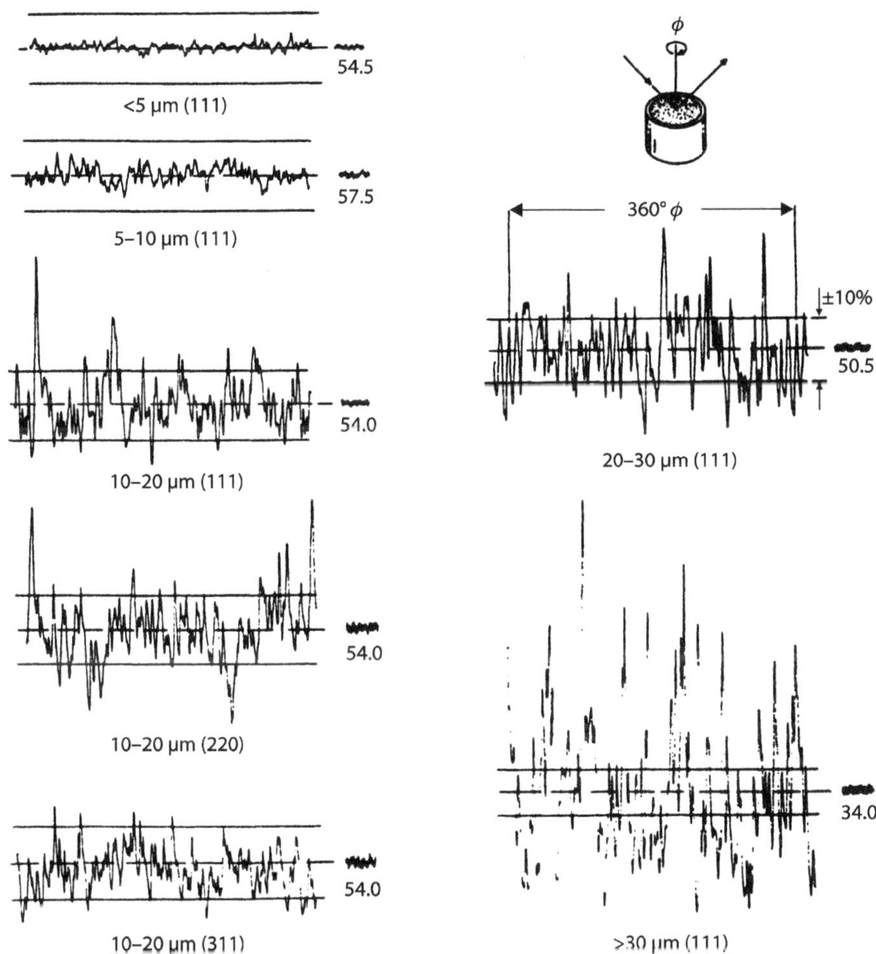

Fig. 3.23. Recordings of intensity variations when rotating the specimen during slow azimuthal rotation. Shown are recordings for several reflections of a silicon specimen measured with Cu Kα X-rays. At the *right side* the recording with the general fast rotation is depicted

pointed out that R is a number, calculated from observed and calculated values. It is neither a parameter nor a factor. A factor is a number in an equation. This is not the case here. So the term "R-factor" as it is sometimes called is wrong and must not be used.

3.6.2
R-Values in Powder Diffraction

When the quality of structural data from powder diffraction was discussed, authors took reference to this well known single crystal R-values, shown in Eq. 3.29. However there is a conflict and occasionally confusion about how to define this values in pow-

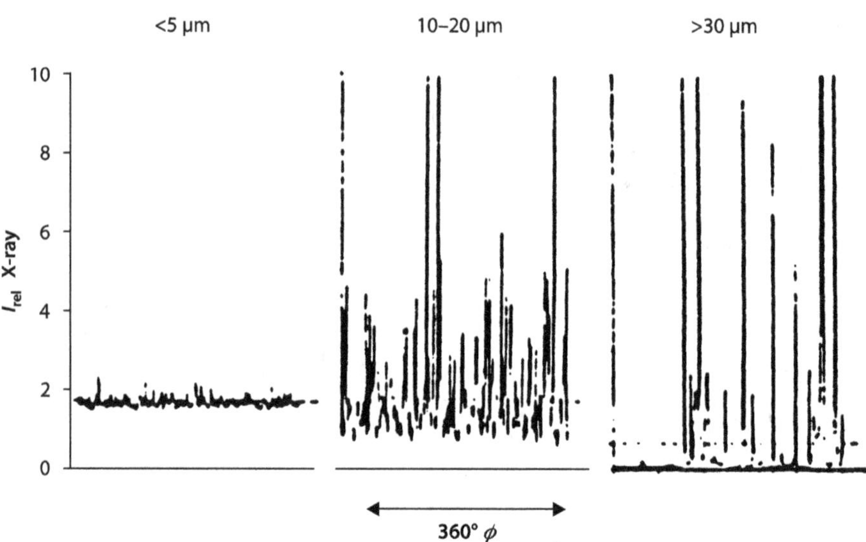

Fig. 3.24. Effect of particle size on the intensity of the Si(111) peak during slow azimuth rotation ϕ, measured with highly parallel storage ring synchrotron X-rays

der diffractometry. In single crystal work the concept is straight forward since structure factors, observed and calculated ones are compared. Using the same definition in step scanning patterns, we have a different situation. In Rietveld calculations the data are y_i(obs) and y_i(calc), observed and calculated counts for each step of a diffraction pattern. Here the number of scanning steps and therefore of observations is huge, hundreds to thousands. Since the number of observations enter the calculation of the R-value in the denominator the R-values, and also the estimated standard deviations of structural parameters are necessarily low. These discrepancies have been recognized quite early and have been discussed repeatedly, for example by Haywood and Shirley (1977), Sakata and Cooper (1979), Scott (1983), Prince (1985), or by Taylor (1987). A detailed discussion of R-values has been presented by Jansen E. et al. (1994a). It must be discussed therefore in more detail.

3.6.3
Background

The main reasons for many discrepancies are found in the treatment of background. Reliable determinations of integrated intensities require a good knowledge of the background. This is especially important with powder patterns, where we may have amorphous contributions. Or in synchrotron diffraction in the energy dispersive mode, where we observe commonly a background which is in general not a smooth line. In good powder diffraction applications the background must be separated from the crystalline contributions before structure refinement or analysis. It influences the quality of the data and the ensuing refinement and most serious it has considerable influence on the calculated R-values, which are taken as a standard for the quality of the analysis. These difficulties and ambiguities do not occur in the Two Stage method, where first integrated intensities are determined by profile fitting. They are then used for the crystallographic calculations. This method approaches therefore single crystal quality.

3.6.4
Dealing with Background

The background in powder diffraction patterns is best described by one or several polynomials. This has been discussed in Sect. 3.3 in connection with profile fitting. In one specific case when analyzing synchrotron energy dispersive diffraction diagrams Lauterjung described the background by a sum of n orthogonal polynomials (Lauterjung et al. 1985). It may be recommended as a general and good approach especially in energy dispersive synchrotron experiments, where the background changes considerable with energy. In these patterns the background is determined by the spectral distribution of the primary beam, by the energy dependent absorption in the specimen (for example when the energy is going through an absorption edge) and by diffuse scattering of the phonons. The background then varies slowly with phonon energy. Lauterjung in his program used a modified formalism of a method given by Steenstrup (Steenstrup 1981) with n polynomials up to order five. Figure 3.25 shows as a representative example the diffraction pattern of the high temperature phase of $MnSO_4$ taken at 460 °C. The variation and adjustment of the background shows up clearly. In this specific case it was described by a polynomial of order five. Because of the closeness of diffraction peak, here as in many other problems, the Rietveld way with footing marks between the peaks would not give acceptable results.

3.6.5
Background in Rietveld Programs

Rietveld in his original version subtracted the background before the actual refinement by defining a polynomial. The count rates y_i which he used in his refinements were therefore above background and the calculated *R*-values referred correctly to measured intensities coming from the sample. The *R*-value was calculated as (Rietveld 1969; Young et al. 1982)

$$R = \frac{\sum_i |y_i(\text{obs}) - y_i(\text{calc})|}{\sum_i |y_i(\text{obs})|} \tag{3.30}$$

Since the first publication of the source code T418/419 by Rietveld (1969), many authors have modified the code of the program in order to achieve a simple operation and universal application. In some modern versions of the Rietveld programs the background is included without prior subtraction directly in the least squares calculations by manually setting footing marks defining areas of background between the diffraction peaks and describing it by a polynomial. These background sections are excluded

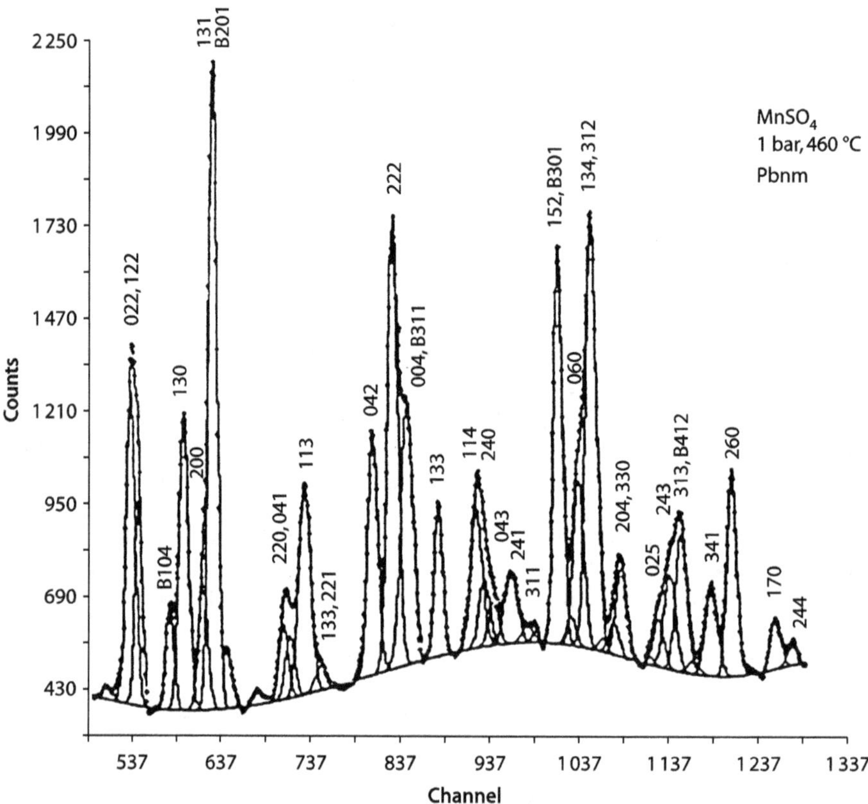

Fig. 3.25. Diffraction pattern of $MnSO_4$ collected with synchrotron X-rays at 460 °C

in the Rietveld refinement. This procedure is still used today in most Rietveld programs. These modifications have often influenced the results, even if in many cases the authors were unaware of them.

We have considered, and tested, two program modifications that we find in some widely used Rietveld versions: The first introduces a more convenient treatment of the background and affects the profile R-values, the second simplifies the estimation of "observed" Bragg intensities, which has consequences for the Bragg R-value.

In the following we discuss these effects for a neutron diffraction pattern that was measured in the course of a structure refinement of $Ba[O(H,D)]Br \cdot 2(H,D)_2O$. The diffraction pattern is shown in Fig. 3.26. It has been refined with a Rietveld program. The differences are included in the box above. This sample, and example, is well suited for our discussion, because we have a moderately high background originating from incoherent scattering of hydrogen. We always used the same model and did not refine any parameters, neither structural nor instrumental. So the only differences are the R-values given by the different programs.

The intensity values, i.e. the count rates used in such refinements and in the calculations of the R-values are then

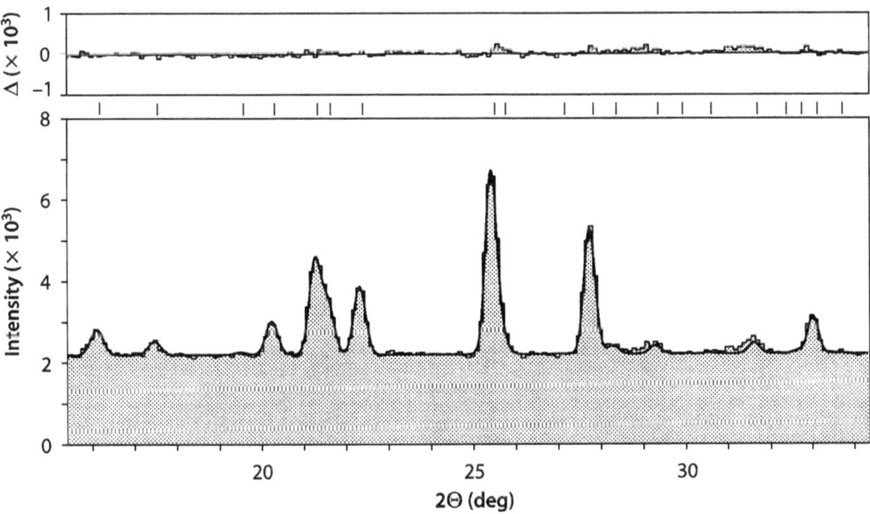

Fig. 3.26. Section of a measured and Rietveld-refined neutron diffraction pattern of Ba[O(H,D)]Br · 2(H,D)$_2$O. The high background originates from incoherent scattering of hydrogen. Calculated reflection positions are given in the middle by marks. Differences between observed and calculated count rates at the *top*

$$y_i'(\text{obs}) = y_i(\text{obs}) + b_i \tag{3.31a}$$

$$y_i'(\text{calc}) = y_i(\text{calc}) + b_i \tag{3.31b}$$

with b_i the estimated background contributions to the pattern, observed as well as calculated. With this definition we arrive at a different, sometimes significantly modified R-value, including background in the denominator:

$$R' = \frac{\sum_i |y_i'(\text{obs}) - y_i'(\text{calc})|}{\sum_i |y_i'(\text{obs})|} \tag{3.32}$$

With poor counting statistics when the ratio signal to noise is low it is not without consequences whether we are using Eq. 3.30 or Eq. 3.32. This arbitrariness and deficiency has been recognized and pointed out quite early by Jansen and co-workers (Jansen E. et al. 1989) and by Hill and Fischer (1990). It can be easily understood if we look at Eq. 3.32: The differences $|y_i'(\text{obs}) - y_i'(\text{calc})|$ in the nominator are the same in both equations, since the background contributions b_i in Eq. 3.32 cancel each other. The denominators however are drastically different in both equations depending on the amount of background. Figures 3.27 and 3.28 demonstrate the calculation with Eqs. 3.30 and 3.32, respectively. The data used are identical. The differences between

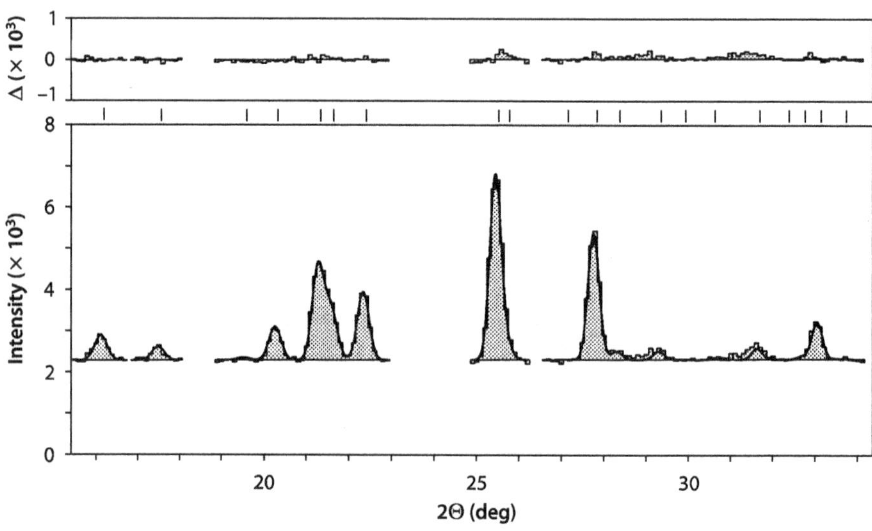

Fig. 3.27. Profile refinement as in Fig. 3.26 where only count rates above background are considered. $R = 19.86\%$

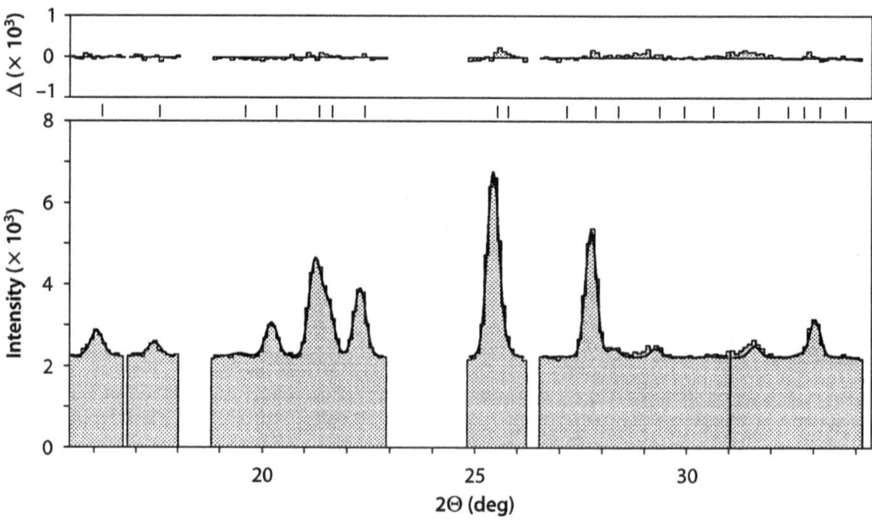

Fig. 3.28. Profile refinement as in Fig. 3.26, but including background. $R = 2.06\%$

observed and calculated count rates, shown in the upper boxes, are identical in both figures. These differences represent the nominator in the two equations. Obviously the result must lead to different R-values. For Fig. 3.28 $R'_{\text{Rietveld}} = 2.06\%$, according to Eq. 3.30. In contrast for Fig. 3.27 and Eq. 3.32 $R'_{\text{Rietveld}} = 19.86\%$, the "true" R-value.

3.6.6
The Profile *R*-Value

We must distinguish *R*-values describing the goodness-of-fit of the profiles, R_{PF}, or $R_{profile}$, or R_p (profile fitting *R*-value), and *R*-values based on the resulting crystallographic parameters, i.e. atomic coordinates etc., obtained through least squares calculations, R_{Bragg}. Unfortunately those values are often mixed up and named just "*R*-value".

The profile *R*-values are calculated (correctly) from the differences between y_i(obs) and y_i(calc) at the observed step-scanning points $\Delta(2\Theta)$, prior to any crystallographic calculation or refinement. It may be calculated for a single peak, for a cluster of peaks or for the entire pattern, like in the program FULFIT. The goodness of the fit is consequently expressed by a profile *R*-value. This profile *R*-value depends essentially on the absolute number of counts y_i(obs). This is reasonable when considering only counting statistics and assuming pure Bragg scattering. In practice however, the observed counts contain a certain amount of background scattering originating from the sample, i.e. incoherent scattering, and from the instrument, for example detector noise. The background adds to the Bragg scattering but of course is independent of it. Consequently, $R_{profile}$-values calculated by Eq. 3.30 depend on nonstructural effects: the higher the background, the better, i.e. the lower the *R*-value $R_{profile}$. Thus it is possible to obtain an excellent agreement index even with a poor diffraction pattern if it is characterized by an unfavorable peak-to-background ratio. The mathematically correct *R*-values have lost their physical meaning.

A further aspect has to be discussed in connection with the total number of data points used for the calculation of *R*-values. In recent versions of Rietveld programs this depends on the profile widths and the treatment of the background. In Rietveld's original program (written for neutron diffraction data) only Gaussian peak profiles were considered. The range around a peak position possibly may still contain Bragg scattering. It is related to the half width, i.e. *FWHM* (full width at half maximum). This is taken into account by a fixed factor in the program, for example ±1.5 times *FWHM*. The *R*-values calculated for Figs. 3.27 and 3.28 are based on this consideration. Later Rietveld versions can handle peak profiles other than Gaussian, for example Lorentzian or Pseudo-Voigt with broader contributing ranges by changing this "width factor". When the width factor is changed, in our example from 1.3 to 3.0 for instance, R_p will increase from 19.86 to 20.85% (Eq. 3.30), and R'_p decrease from 2.06 to 1.99% (Eq. 3.32). When the automatic background fitting facilities of modern program versions are used, all point are included in the refinement and consequently also in the *R*-value calculation. This is the shaded area in Fig. 3.27. This diagram with data based on all data points leads to a calculated profile *R*-value R_p = 21.63% (without background), and R'_p = 1.93% (including background counts). This difference demonstrates how published *R*-values have to be looked at with some suspicion.

In the denominator, large numbers yield automatically low *R*-values, and such values are therefore misleading. It has been recognized and discussed, for example by Haywood and Shirley (1977), Sakata and Cooper (1979), Scott (1983), Prince (1985), or Taylor (1987). The other problem is the inclusion, or omission, of the background. A high background will automatically lead to small differences between observed and calculated values and again to a misleading low R-value. Both deficiencies therefore lead often to severe deviations from the true values.

3.6.7
The Weighted Profile R-Value

The function to be minimized in the least squares profile refinement routine is

$$M = \left(\sum_i w_i \left(y_i(obs) - y_i(calc) \right)^2 \right)^{1/2} \tag{3.33}$$

with w_i the weights taken from the experimental error margins. This is usually the Poisson counting statistics taken as $w_i = 1 / y_i'(obs)$. It should be emphasized, that the experimental standard deviations of the observed counts $y_i'(obs)$ and thus the weight parameter w_i depend on both Bragg scattering and background. Accordingly, the agreement index of the profile fit without background is then expressed by

$$R_{wp} = \frac{\sqrt{\sum_i w_i \left(y_i(obs) - y_i(calc) \right)^2}}{\sqrt{\sum_i w_i y_i(obs)^2}} \tag{3.34a}$$

or, when the background is (wrongly) included in the counts

$$R_{wp}' = \frac{\sqrt{\sum_i w_i' \left(y_i'(obs) - y_i'(calc) \right)^2}}{\sqrt{\sum_i w_i' y_i'(obs)^2}} \tag{3.34b}$$

Another value often calculated is the so-called "expected profile R-value" R_{exp}, which indicates a possibly attainable limit for R_{wp} by considering only pure statistics in the nominators of Eq. 3.34. With n the number of steps, i.e. the intensity values, and m the number of refined parameters, the definitions are for R_{exp} after background subtraction and R_{exp}' including background in the observed intensities.

$$R_{exp} = \frac{\sqrt{n-m}}{\sqrt{\sum_i w_i y_i(obs)^2}} \tag{3.35a}$$

$$R'_{\text{exp}} = \frac{\sqrt{n-m}}{\sqrt{\sum_i w_i' \, y_i'(\text{obs})^2}}$$ (3.35b)

A summary of the profile *R*-values obtained with the test example is given in Table 3.8. It reveals very convincingly the strong effect of background. In order to avoid unreasonable values in *R* and to exclude nonstructural effects, the calculation of profile *R*-values, R_{PF}, should be performed with the pure Bragg scattering data $y_i(\text{obs})$, i.e. after subtraction of background. This is the case by definition when the refinement is performed with POWLS in the Two Stage method.

3.6.8
The Bragg *R*-Value

Judging the quality of the structure refinement and the obtained crystal parameters, especially atomic positional parameters, temperature factors, occupation factors etc., the so-called Bragg *R*-value R_{Bragg} is the important value, and this value is more important than the profile *R*-value. The Two Stage method and the program POWLS calculates only Bragg *R*-values. A real value R_{Bragg} is based on integrated intensities and is defined accordingly to

$$R_{Bragg} = \frac{\sum_j \left| I_j(\text{obs}) - I_j(\text{calc}) \right|}{\sum_j \left| I_j(\text{obs}) \right|}$$ (3.36)

I(obs) and *I*(calc) are respectively observed and calculated integrated Bragg intensities (without background). The summations run over all Bragg reflections. It must be remembered that in Rietveld refinements values *I*(obs) are not available; they are not necessary for the refinement procedure because model and observations are

Table 3.8. Profile *R*-values (in %) obtained by Rietveld refinement of the neutron diffraction pattern of Ba[O(H,D)]Br·2(H,D)$_2$O shown in Fig. 3.26. Shown is also the influence of the integration width

	Definition	Integration width				Definition	Integration width		
		±1.5 *FWHM*	±3.0 *FWHM*	No limits			±1.5 *FWHM*	±3.0 *FWHM*	No limits
R_{profile}	Eq. 3.30	19.86	20.85	21.63	R'_{profile}	Eq. 3.32	2.06	1.99	1.93
R_{profile}	Eq. 3.34a	14.28	14.56	14.73	R'_{profile}	Eq. 3.34b	2.71	2.64	2.57
R_{exp}	Eq. 3.35a	10.41	10.96	11.44	R'_{exp}	Eq. 3.35b	1.98	1.98	1.99

compared directly on the basis of step intensities. Nevertheless all Rietveld program versions supply an agreement index R_{Bragg}.

The estimation of observed integral intensities in overlapping peaks in Rietveld programs is done by decomposing the observed counts $y_i(\text{obs})$ according to the *shares* of the calculated Bragg intensities $I(\text{calc})$ and distributing them in this ratio as the $y_i(\text{calc})$ contribute to the corresponding $I(\text{calc})$ values. Each $y_i(\text{calc})$ then consists of contributions $p_{i,j}(\text{calc})$ according to

$$y_i(\text{calc}) = \sum_j p_{i,j}(\text{calc}) \tag{3.37}$$

Figure 3.29 shows the composition of the pattern in our example. The shaded (Gaussian) areas represent Bragg intensities, which make up the diagram. Because of overlap some Gaussians look distorted. The contributions $p_{i,j}(\text{calc})$ in Eq. 3.37 are shares of Bragg intensities $I_j(\text{calc})$:

$$y_j(\text{calc}) = \sum_i p_{i,j}(\text{calc}) \tag{3.38}$$

Analogous to this summation, the so-called "observed" integrated Bragg intensities $I(\text{obs})$ in the Rietveld formalism are calculated accordingly to

$$I_j(\text{obs}) = \sum_i p_{i,j}(\text{obs}) \tag{3.39}$$

In order to obtain $p_{i,j}(\text{obs})$, a proportionality relation is assumed:

$$p_{i,j}(\text{obs}) = \frac{y_i(\text{obs})}{y_i(\text{calc})} p_{i,j}(\text{calc}) \tag{3.40}$$

The Bragg R-value R_{Bragg} according to Rietveld's original version is then

$$R_{\text{Bragg}} = \frac{\sum_j \left| \sum_i \frac{y_i(\text{obs})}{y_i(\text{calc})} p_{i,j}(\text{calc}) - I_j(\text{calc}) \right|}{\sum_j \left| \sum_i \frac{y_i(\text{obs})}{y_i(\text{calc})} p_{i,j}(\text{calc}) \right|} \tag{3.41}$$

Figure 3.30 depicts a graphic visualization of the Rietveld "observed" and calculated intensities which were used for the calculation of R_{Bragg} according to Eq. 3.41. For this

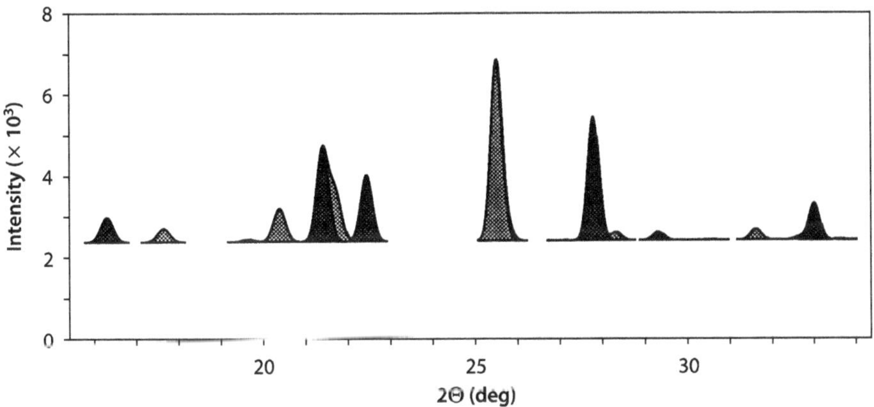

Fig. 3.29. Composition of the calculated pattern of Fig. 3.26 from the calculated Bragg intensities, spread over Gaussian peak shapes. Some peaks are partly overlapped and the Gaussian shapes look therefore deformed

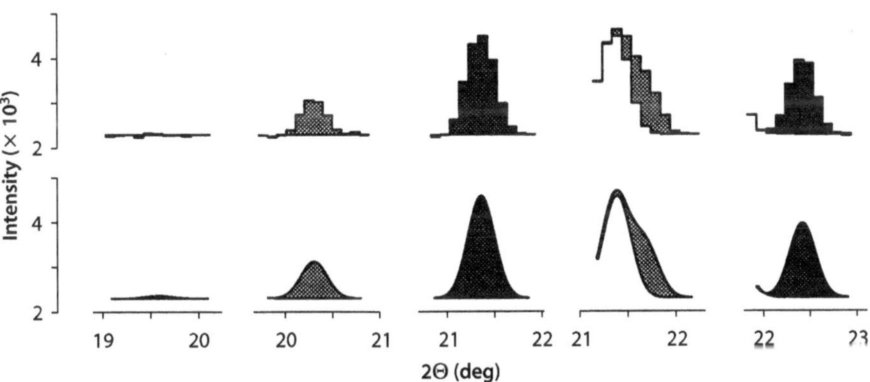

Fig. 3.30. Representation of "observed" and calculated Bragg intensities in the cluster between 19 and 23° of a diffraction pattern measured with neutrons. *Shaded areas* show the "observed" (*top*) and calculated intensities (*bottom*), which determine the *R*-value R_{Bragg} of Eq. 3.41

estimation of $I(\mathrm{obs})$ it is necessary to store the values $p_{i,j}(\mathrm{obs})$ in the course of the calculation.

Difficulties and discrepancies arise when some program versions try to obtain the Bragg intensities in a simplified manner. By analyzing the source code of such programs one finds that the evaluation of the "observed" intensities $I'(\mathrm{obs})$ is done on the basis of integrated step intensities under the assumption that the relation of Eq. 3.42 is valid:

$$I_j{}'(\mathrm{obs}) = \frac{\sum\limits_{k(j)} y_k{}'(\mathrm{obs})}{\sum\limits_{k(j)} y_k{}'(\mathrm{calc})} I_j(\mathrm{calc}) \tag{3.42}$$

The summation index k here runs over all measured data points that can contribute to the reflection j. When $I_i'(\text{obs})$ is introduced into Eq. 3.36 again another Bragg R-value R'_{Bragg} is obtained, different from that of Eq. 3.43.

$$R'_{\text{Bragg}} = \frac{\sum_j \left| \dfrac{\sum_k y_k'(\text{obs})}{\sum_k y_k'(\text{calc})} I_j(\text{calc}) - I_j(\text{calc}) \right|}{\sum_j \left| \dfrac{\sum_k y_k'(\text{obs})}{\sum_k y_k'(\text{calc})} I_j(\text{calc}) \right|} \qquad (3.43)$$

Figure 3.31 depicts the areas of $\Sigma y_i'(\text{obs})$ and $\Sigma y_i'(\text{calc})$, the areas which were used in Eq. 3.43. It is obvious that these ratios tend to become 1 with increasing background and widths factors; consequently, R'_{Bragg} tends to become zero. R'_{Bragg} is identical to R_{Bragg} in Eq. 3.41 if and only if

i   the background vanishes,
ii   the reflections do not overlap, and further
iii the width factors are identical.

The tremendous difference between R'_{Bragg} and R_{Bragg} is shown in Table 3.9.

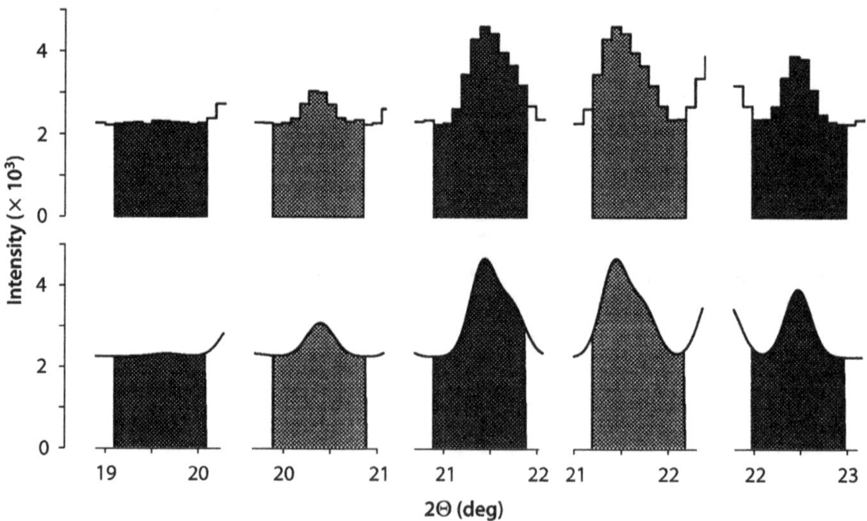

Fig. 3.31. Representation of observed and integrated count rates in the cluster between 19 and 23°, similar to Fig. 3.30. These areas determine the R-value R'_{Bragg} of Eq. 3.43

Table 3.9. Dependence of R_{prof} on background scattering (see also Fig. 3.32)

Test refinements		R_{prof} (%)	R_{prof} (%)
Example	Background	Eq. 3.30	Eq. 3.31
Fig. 3.32a	Medium (original)	2.6	14.9
Fig. 3.32b	High	1.5	13.5
Fig. 3.32c	Small	12.7	23.5

3.6.9
Conclusion

From this discussion we have to conclude:

- The profile R-values R_{p}, R_{wp}, R_{exp} are of significance with respect to structural refinements only if they are calculated from pure Bragg scattering, i.e. from background cleaned intensities.
- Bragg R-values R_{Bragg} are highly dependent on the procedure in which "observed" integrated intensities are estimated during the course of the program. When evaluated on the basis of integrated counts, they are often artifacts and do not have the meaning of R-values we know from single crystal data or from Two-Stage refinements.
- These problems do not occur in the Two Stage method, since the program POWLS is based on integrated intensity observations above background. Just like in single crystal work. The R-values calculated in POWLS are therefore directly comparable to single crystal R-values.

Finally to demonstrate the dangerous influence of background on the R-values Fig. 3.32 shows the neutron diffraction patterns of $Ba[O(H,D)]Br \cdot 2(H,D)_2O$. In this compilation Fig. 3.32a depicts the original, measured diffraction pattern with a value $R = 2.6\%$. In Fig. 3.32b 2 000 counts have been added artificially to the background resulting in a lower value of $R = 1.5\%$. In Fig. 3.32c 2 000 counts have been subtracted, thus reducing the background. This last one, after subtracting the background, would be the proper one and the profile R-value would be only 12.7%, in contrast to Fig. 3.34b, where with a high background we have the totally misleading R-value of 1.5%. Table 3.9 summarizes this study.

3.7
Structure Refinement by the Two Stage Method

3.7.1
Refinement with POWLS

The Two Stage method with its origin in neutron diffraction has shown to be superior to total-pattern fitting methods if we are confronted with unconventional structures, unconventional diffraction techniques or environments with contaminating peaks.

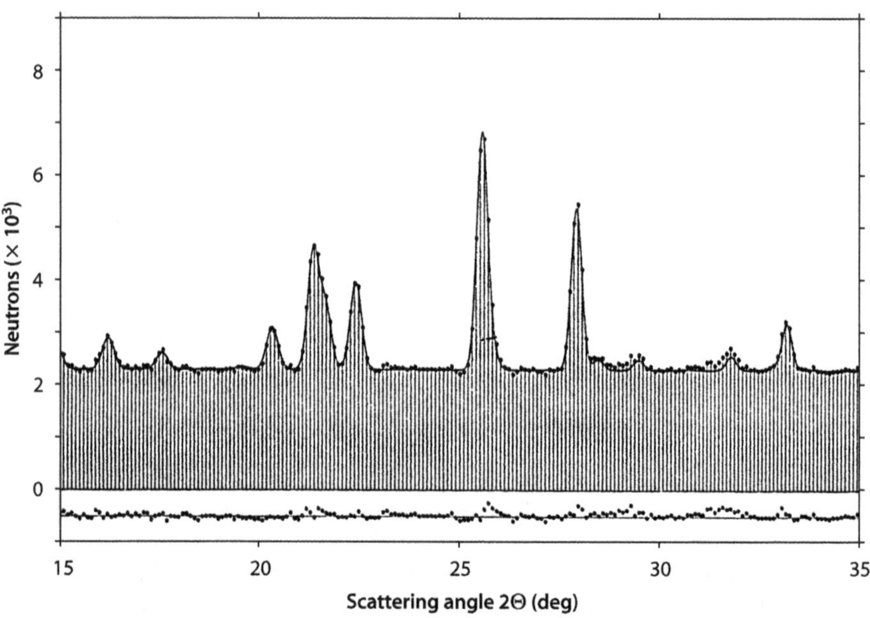

Fig. 3.32 a. Neutron diffraction pattern of Ba[O(H,D)]Br · 2(H,D)$_2$O with different backgrounds; original measurement, $R_{prof} = 2.6\%$

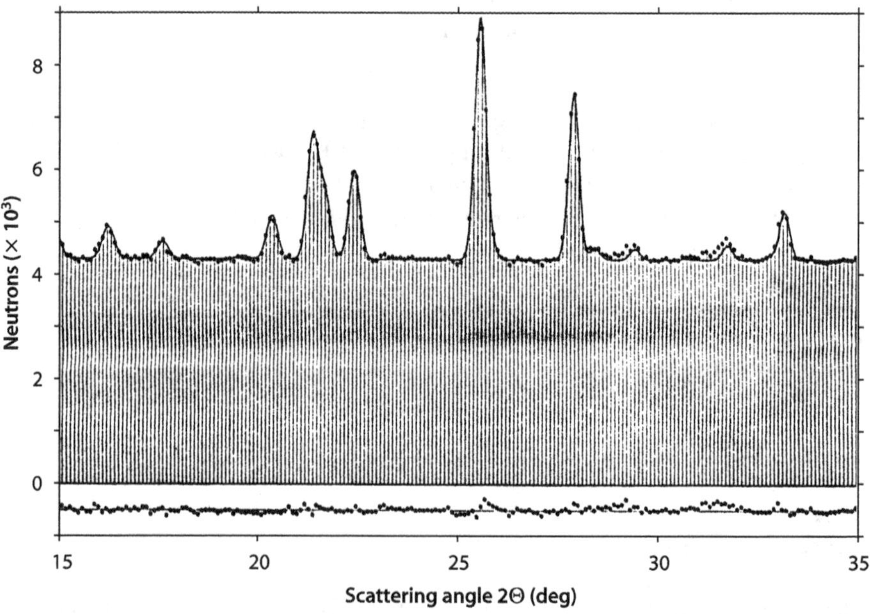

Fig. 3.32 b. Neutron diffraction pattern of Ba[O(H,D)]Br · 2(H,D)$_2$O with different backgrounds; 2000 counts added, $R_{prof} = 1.5\%$

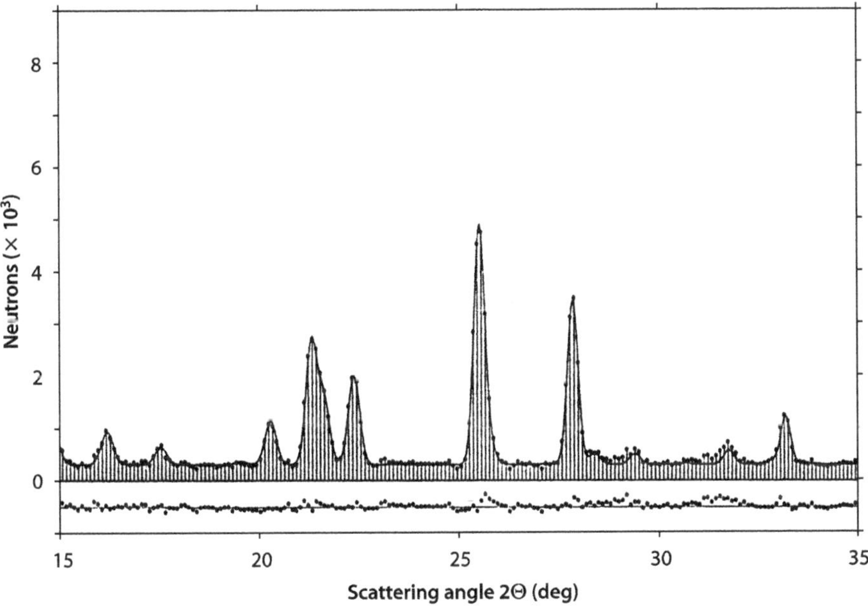

Fig. 3.32 c. Neutron diffraction pattern of $Ba[O(H,D)]Br \cdot 2(H,D)_2O$ with different backgrounds; 2 000 counts subtracted, $R_{prof}=12.7\%$

Such cases are found in neutron diffraction in the determination of magnetic structures. If those structures are non-commensurate or have a helical moment arrangement which shows up in satellites, often very close to the main peak, it is not possible to refine the structure by a full pattern approach without very extensive programming adapted to the (likely) structure. Even pattern decomposition need extensive interaction by the operator. Another uncommon case is found when position sensitive detectors are used, and this is especially true in neutron diffraction, where there are commonly deviations from linearity. Other examples are found in cases with in situ high pressure X-ray diffraction where there are always maxima from the sample holder, from the pressure marker and from other containments. Such a case is shown in Fig. 3.9. Also in the study of phase transformations under pressure (or also temperature) two phases are in general present at the beginning of the phase transformation, which must be separated for further analytical calculations, for example to calculate lattice parameters.

The Two Stage method has been extended since its beginning in 1962 by detailed mathematical pattern decomposition procedures, as has been discussed in Sect. 3.3, and also in Sect. 1.2. The profiles may be complicated and therefore must be analyzed first on a standard material, like LaB_6, which has been recommended for X-ray diffraction. This is followed by the actual profile-fitting procedure. The structural parameters are then refined separately from instrumental variables, because profile analysis as well as profile fitting are independent of structural informations. The profile fitting or pattern decomposition yields a list of integrated intensities besides peak positions and halfwidth. The structural parameters are refined in a manner analogous to single crystal structure determination. The number of observations in the Two Stage method is fixed at the (*hkl*) population. From this list of data the structure-relevant information can be sorted out. This means only the peaks belonging to the structure in question are used for further analysis.

For the decomposition of the patterns several programs are available worldwide, like the one published by Pawley (1981). In our own working groups three programs were developed and used: PROFAN (Merz et al. 1990), FULFIT (Jansen E. et al. 1988), and HFIT (Will and Höffner 1992; Höffner and Will 1991) designed specifically for high pressure synchrotron data. The crystal structure refinement program POWLS-80 (Will 1979; Will et al. 1983b) has been used in numerous applications. It is well documented and the source code is included in the Appendix.

3.7.2
Examples Using the Two Stage Method and the Program POWLS

Determination of Incommensurate Magnetic Structures

For the determination of incommensurate magnetic structures the Rietveld method in general fails completely. Only separating the peaks will lead to a solution. This is true already if we are dealing with conventional linear magnetic structures, where however in most cases the magnetic unit cell differs from the crystallographic unit cell. If we want to use a Rietveld type program we do need a magnetic model first, and also extensive modifications of the Rietveld routine, for example to calculate the magnitude and direction of the magnetic moments. It becomes exceedingly difficult if we are dealing with helical structures, where the moment arrangement manifests itself through satellite peaks. Depending on the size of the helix, i.e. the unit repeat on rotation of the magnetic moments, the satellites may be very close to the main peak, and more over there may be more than one satellite. Such an example is $TbAsO_4$ where there is a very complex magnetic helical structure below 1.5 K (Schäfer et al. 1979; Kockelmann et al. 1992a,b). Figure 3.33 shows in part the low temperature neutron diffraction pattern with magnetic satellites. The pattern is characterized by a complex scheme of seven satellite reflection, which cannot be interpreted by a simple spiral. The successful profile analysis has been performed interactively with PROFAN by adding one more satellite after the other which showed up in the difference diagrams, after a complete satisfactorily fitting was reached. The final splitting, and profile fitting, was then performed using the program FULFIT, which puts the corresponding satellites at each main (nuclear) peak position (Fig. 3.33).

3.7.3
Comparison of the Rietveld Method and the Two Stage Method

How does the Rietveld method and the Two Stage method compare. If the applications are strictly refinements of crystal structures both must necessarily yield the same results. A careful comparison is therefore useful. The present discussion is based on the same experimental data. Available were two diffraction patterns of quartz, both measured with synchrotron radiation at different wavelength and 1 year apart. This

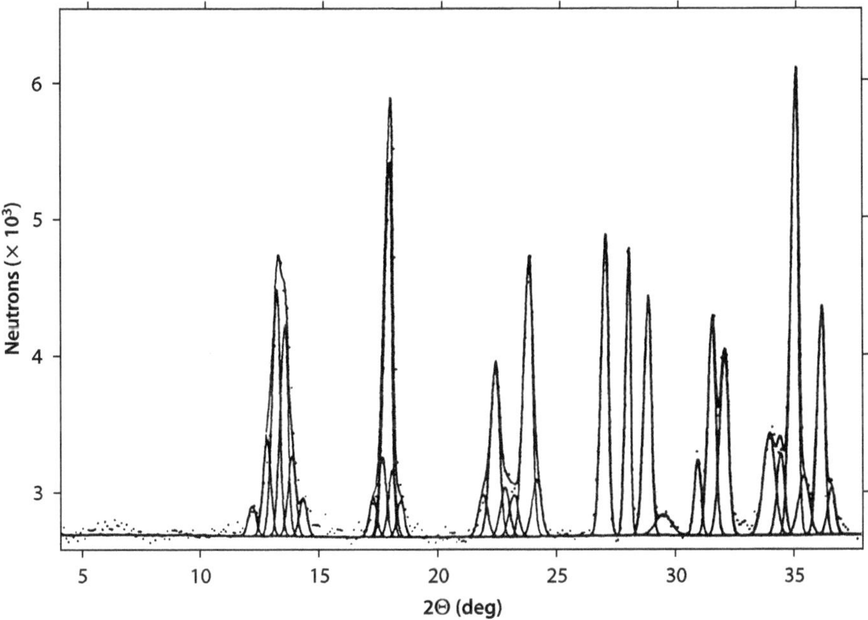

Fig. 3.33. Low-temperature neutron diffraction pattern of TbAsO$_4$ with magnetic satellite splitting analyzed by FULFIT (*lines*)

comparison refers only to structure refinements of the crystallographic parameters, atomic positions and Debye-Waller factors. The results should come out the same, if the two methods are applied properly.

Both data sets have been analyzed by the Two Stage method using the least squares program POWLS; and in an identical way by the Rietveld full pattern method. For the Rietveld refinement the program version DBW3.2 (Wiles 1982a) was used. It is well documented and needs no further description.

The first data set was measured with $\lambda = 1.0020$ Å, step size $\Delta(2\Theta) = 0.02°$ with an angular range of $2\Theta = 45$ to $135°$. To explore the influence of the size of experimental data, that means the range of data in reciprocal space, the patterns were analyzed in 3 sections. The first section of the first pattern ranged from $2\Theta = 45$ to $75°$ ($s = 0.60$ Å^{-1}) with 31 peaks coming from 47 sometimes intrinsically overlapping planes. The program POWLS has provisions to calculate intrinsically overlapping peaks individually and include them in the refinement. The resulting parameters are listed in Table 3.10. Refinement was then extended to data from $2\Theta = 45$ to $105°$ ($s = 0.79$ Å^{-1}), and finally to the full data set up to $2\Theta = 135°$ ($s = 0.92$ Å^{-1}) with 157 resolved peaks from 280 Miller planes. All results are shown in Table 3.10. They are in excellent agreement with a single crystal study by Levien et al. (1980), who used 210 planes (see Table 3.15). The standard deviations from the powder data are about three times higher than from the single crystal data, which has to be expected.

The second data set was measured with $\lambda = 1.2823$ from $2\Theta = 15$ to $115°$ ($s_{max} = 0.66$ Å^{-1}) one year later but with the same specimen. In this case a smaller step width of

Table 3.10. Positional parameters, temperature factors B and R-values of quartz refinement calculations performed with the Two Stage and the Rietveld method in comparison for different angular range extensions. Two patterns measured at different wavelength are analyzed

	FULFIT/ POWLS	Rietveld	FULFIT/ POWLS	Rietveld	FULFIT/ POWLS	Rietveld
λ (Å)	2 Theta-sections (deg)					
1.0020	45–135		45–105		45–75	
x(Si)	0.4701(2)	0.4699(1)	0.4701(3)	0.4698(1)	0.4703(3)	0.4700(1)
x(O)	0.4106(7)	0.4118(3)	0.4105(8)	0.4119(3)	0.410(1)	0.4111(5)
y(O)	0.2658(4)	0.2666(2)	0.2658(6)	0.2666(2)	0.2657(7)	0.2665(4)
z(O)	0.2848(4)	0.2853(1)	0.2847(5)	0.2853(1)	0.2850(5)	0.2854(2)
B(Si)	0.45(1)	0.470(5)	0.44(2)	0.464(7)	0.51(4)	0.54(2)
B(O)	0.96(3)	0.96(1)	0.97(5)	0.97(2)	1.06(8)	1.07(4)
R_{prof}	0.0360	0.0459	0.0406	0.0504	0.0438	0.0524
R_{Bragg}	0.0446	0.0320	0.0402	0.0297	0.0248	0.0172
λ (Å)	2 Theta-sections (deg)					
1.2823	15–115		15–90		15–60	
x(Si)	0.4704(5)	0.4702(1)	0.4704(5)	0.4702(1)	0.471(1)	0.4708(1)
x(O)	0.4131(6)	0.4131(2)	0.4131(8)	0.4132(2)	0.413(1)	0.4133(3)
y(O)	0.2668(8)	0.2669(2)	0.2667(8)	0.2668(2)	0.267(1)	0.2675(3)
z(O)	0.2853(6)	0.2863(1)	0.2853(7)	0.2865(2)	0.285(1)	0.2875(3)
B(Si)	0.21(2)	0.281(6)	0.21(4)	0.281(6)	0.2(1)	0.33(2)
B(O)	0.68(6)	0.63(2)	0.65(9)	0.61(2)	0.2(2)	−0.03(5)
R_{prof}	0.0347	0.1029	0.0361	0.1012	0.0218	0.1058
R_{Bragg}	0.0241	0.0263	0.0211	0.0259	0.0122	0.0189

$\Delta(2\Theta) = 0.01°$ was used together with improved experimental conditions, namely a longer wavelength, a smaller step width, and also a narrower collimator system giving much improved spatial resolution in the pattern. The halfwidth was now $0.08°$. This resulted in 72 well resolved peaks coming from 119 Miller planes. All peaks could be separated very well by profile fitting. It is worth to note the intensity range of the peaks: The peak height of the strongest peak, (101)/(011) was 35 494 counts, the weakest peak, (222), had a peak height of only 5 counts. This is an intensity range of 7 000 to 1. The integrated intensities together with the calculated values are listed in Table 3.11. The individual profile R-values obtained with PROFAN are in the range 1.4 to 2.0%.

Also here the refinement calculations with POWLS were performed in three sections, beginning with a smaller, lower angle data set of $2\Theta = 15$ to $60°$ with the first 60 peaks coming from 150 *hkl* planes, then $2\Theta = 15$ to $90°$ and finally the full data set $2\Theta = 15$ to $115°$. Interesting the best results were obtained with 40 observations from 64 *hkl* planes, which resulted in $R = 1.41\%$. (Not included in Table 3.10) With 8 variables the system is already 5-fold overdetermined, therefore the limitation of the data set is well justi-

Table 3.11. Comparison of observed and calculated intensity data for quartz determined by FULFIT profile fitting and POWLS refinement

hkl	I(obs)	I(calc)	hkl	I(obs)	I(calc)	hkl	I(obs)	I(calc)
100	7409	7161	311,131	1281	1263	106,016	119	109
101,011	35494	35834	204,024	146	163	412,142	494	433
110	3018	2995	222	5	18	305,035	116	128
102,012	2844	2750	303,033	130	182	323,233	8	12
111	1527	1345	312,132	1928	1940	500	1	0
200	2242	2341	400	333	368	116	148	120
201,021	1559	1409	105,015	803	839	501,051	87	73
112	5987	5776	401,041	615	690	404,044	80	61
003	121	120	214,124	980	999	206,026	612	660
202,022	1891	1837	223	868	904	413,143	786	789
130,013	723	736	402,042	175	182	330	282	248
210	106	119	115	551	559	502,052	696	716
211,121	5032	5024	313,133	385	387	225	300	291
113	879	886	304,034	215	200	331	299	304
300	269	257	320	492	496	420	614	587
212,122	3349	3304	205,025	37	26	315,135	375	333
203,023	3811	3775	321,231	1105	1050	421,241	206	213
301,031	2498	2505	410	98	57	324,234	1034	1022
104,014	1175	1202	322,232	396	374	216,126	313	325
302,032	1797	1784	403,043	704	749	332	16	11
220	1021	1004	411,141	910	926	422,242	484	507
213,123	1918	1830	224	322	312	503,053	8	10
221	534	535	006	104	82	414,144	420	449
114	1474	1450	215,125	591	620	510	173	169
310	1808	1826	314,134	534	544			

fied. With the full data set up to $2\Theta = 115°$ we have 72 observations and 119 Miller planes and the system is 15 fold overdetermined. The resulting R-value is $R = 1.68\%$.

The reason to collect the first data set with a short wavelength and out to almost the limit of detection was done in the hope to improve the standard deviations in the least squares calculations of the parameters by more observations. This turned out to be not so, as can be seen from Table 3.10. The positional parameters did not change or improve outside the error margins when more observations were included in the refinement. From the refinement of data sets in different and increasing segments it can be concluded that intensities up to $s = 0.7$ Å^{-1} equivalent to $2\Theta = 90°$ are sufficient to reach good and reliable results. Remember the full range of the Cu-sphere is $s = 0.65$ Å^{-1}. Limitations are important if one has to consider the time allocated at a synchrotron for the experiment. This is to a lesser degree also true for laboratory measurements.

Figure 3.34 depicts a selected part of the quartz diffraction pattern at high angles after refinements by FULFIT and by Rietveld. As can be realized from the figures and

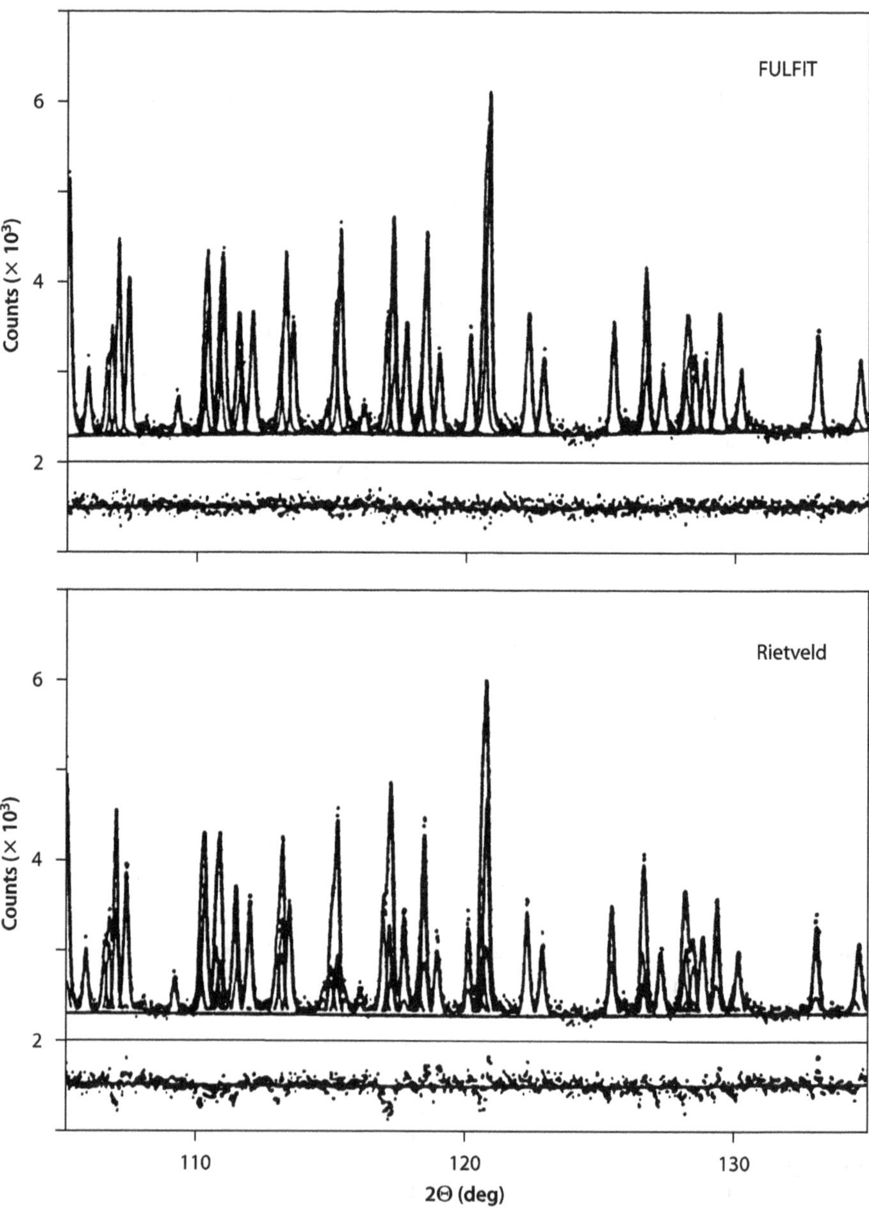

Fig. 3.34. Selected part of a quartz diffraction pattern measured with synchrotron X-rays. The *upper picture* shows the results from profile fitting with FULFIT, the *lower picture* the results from the Rietveld refinement ($\lambda = 1.0020$ Å)

from Table 3.10 the quality is about the same and also the parameters are the same within standard deviations. Both methods give the same results if correctly used. The structure R-values, R_{Bragg}, also differ only slightly. The small differences are likely due to the different handling of the background.

For the Debye- Waller factors on the other hand the results are improving if reflections of small d-values are included, i.e. if high angle data are included. This is understandable, if we remember that bonding effects contribute mainly to the reflections at low diffraction angles and are negligible at higher angles. This is known and used in deformation density studies, where the data sets are split into high order HO and low order LO reflections. Powder diffraction is not the proper means to determine thermal parameters.

Table 3.11 lists observed and calculated intensities derived by FULFIT, the profile fitting program, and POWLS, the least squares refinement program. It seems necessary to comment, that those are real observed intensities, directly comparable to single crystal data. And only such intensities can be used for so-called *ab initio* structure determinations. Intensities given after Rietveld refinements are artificial values, since in overlapping peaks they are artificially divided.

In addition to comparing X-ray diffraction patterns some selected neutron diffraction data collected at the Jülich Research reactor DIDO are included in this comparison. Table 3.12 summarizes crystal structure data from three such investigations, re-

Table 3.12. Crystal structural data from three neutron diffraction measurements refined with the Rietveld program (R) and with POWLS (P)

Space group		Li_2CuO_2 Immm	UFe_4Al_8 I4/mmm	$ThFe_4Al_8$ I4/mmm	$TbNiC_2$ Amm2
Site occupancy		Li in 4(i)	U in 2(a)	Th in 2(a)	Tb in 2(a)
		Cu in 2(a)	Fe in 8(f)	Fe in 8(f)	Ni in 2(b)
		O in 4(j)	Al1 in 8(i)	Al1 in 8(i)	C in 4(e)
			Al2 in 8(i)	Al2 in 8(j)	
Free positions		$z(Li)$	$x(Al1)$	$x(Al1)$	$z(Ni)$
	(R)	0.7144(9)	0.343(1)	0.3488(9)	0.612(1)
	(P)	0.712(1)	0.338(3)	0.351(9)	0.605(2)
		$z(O)$	$x(Al2)$	$x(Al2)$	$y(C)$
	(R)	0.6420(3)	0.281(1)	0.2790(9)	0.151(1)
	(P)	0.6409(7)	0.279(2)	0.273(7)	0.155(2)
					$z(C)$
	(R)				0.302(1)
	(P)				0.298(2)
Number of hkl		70	138	145	70
R_{Bragg}	(R)	0.082	0.076	0.063	0.097
	(P)	0.053	0.084	0.115	0.072

fined both with the Rietveld program DBW3.2 and with POWLS. The table shows good agreement between the two methods. In practice the decision to use one method or the other depends strongly on the goal of the investigation and also on the quality of the diffraction pattern. The Rietveld method turns out to be superior in cases of structures with strongly overlapping peak clusters because of the intrinsic structural constraints, and also if the same, or a similar compound is studied. Such an example is shown in a series of structure refinements of the superconducting 1-2-3 compound, $YBa_2Cu_3O_{6.9}$, which has been studied in a series of several temperatures $T = 16$ K, 82 K, 92 K, 101 K, 183 K and 300 K (Schäfer et al. 1988). In this series the Rietveld routine provides a straight forward strategy, since always the same parameters are included in the refinement. The variation of the cell parameters was here of specific interest, and with orthorhombic symmetry this was a straight forward refinement. Table 3.13 summarizes the results from this series.

3.7.4
Merits and Limitations of the Rietveld Method and the Two Stage Method

The merits of Rietveld's method are found in straight forward conventional structure refinements of common diffraction patterns. This holds

- if we are dealing with single phase substances
- if a model of the crystal structure in question is available
- if the background is simple and easy to be described
- if the profile function is known
- if the unit cell metric (and symmetry) is known approximately

The last two requirements: known profile function and unit cell metric, are conditions *sine qua non*. In the total pattern or Rietveld refinement, a model must be presented to the program, and this model then describes every data point in the diffraction pattern. Without question this is very useful if the pattern is loaded with peaks and it becomes especially true if we are dealing with low symmetry compounds or large unit cells. In both cases we have many peaks close to each other which are sometimes difficult to separate by profile fitting methods. On the other hand in cases of higher symmetry we have large sections in the pattern with a great number of unnecessary data points. This is "ballast" with no contribution to the structure. Also in these examples, many peaks very close together, the definition of background is a serious problem and is one deficiency of the full pattern refinement.

The two methods, Rietveld and Two Stage profile decomposition, are fundamentally different in the case of determining structural parameters. Both are based on a least squares calculation. In the Two Stage refinement the data are the set of decomposed intensities, which resemble a set of single crystal data. This should not be completely artificial if the decomposition process is effective. The Two Stage method should approach the single crystal set as the refinement nears completion. The Rietveld data are a set of step-scan observations y_i, which is quite removed from a single crystal data set. A good proportion of data is concerned with a few well-resolved (and usually more coordinate sensitive) low angle reflections. It is therefore not necessarily true that the error distribution is similar in the two cases and it is also possible that the weighting

Table 3.13. Crystal structural data for the 1-2-3 superconducting compound $YBa_2Cu_3O_{6.9}$: atomic positions, isotropic temperature factors, lattice parameters a, b, c, halfwidth parameters U, V, W, zero-shift of the 2Θ scale and R-values

	$T = 16$ K	$T = 82$ K	$T = 92$ K	$T = 101$ K	$T = 183$ K	$T = 300$ K
z(Ba)	0.1853(5)	0.1856(5)	0.1849(5)	0.1851(5)	0.1850(5)	0.1850(5)
z(Cu2)	0.3550(4)	0.3551(4)	0.3554(3)	0.3549(3)	0.3547(3)	0.3546(3)
z(O2)	0.3792(6)	0.3783(5)	0.3782(6)	0.3775(5)	0.3782(6)	0.3787(6)
z(O3)	0.3752(7)	0.3766(6)	0.3764(7)	0.3781(6)	0.3768(6)	0.3772(6)
z(O4)	0.1594(5)	0.1590(4)	0.1594(5)	0.1602(4)	0.1606(5)	0.1604(5)
B(Y)	0.22(10)	0.25(8)	0.23(9)	0.16(8)	0.19(8)	0.01(8)
B(Ba)	−0.06(11)	0.01(9)	0.06(10)	0.19(9)	0.18(10)	0.42(10)
B(Cu1)	0.34(12)	0.32(9)	0.25(9)	0.19(9)	0.44(10)	0.61(10)
B(Cu2)	−0.15(7)	−0.17(6)	−0.03(6)	−0.13(6)	−0.06(6)	0.00(7)
B(O1)	0.68(19)	0.37(14)	0.45(16)	0.38(15)	0.58(16)	0.81(17)
B(O2)	0.38(11)	0.21(9)	0.39(10)	0.21(9)	0.17(9)	0.36(9)
B(O3)	−0.04(10)	0.14(10)	0.11(11)	0.16(10)	0.11(10)	0.37(11)
B(O4)	0.26(12)	0.32(10)	0.21(10)	0.20(10)	0.36(11)	0.60(11)
a	3.8144(2)	3.8145(3)	3.8149(3)	3.8155(3)	3.8181(3)	3.8226(2)
b	3.8809(3)	3.8806(3)	3.8809(3)	3.8808(3)	3.8825(3)	3.8863(3)
c	11.614(1)	11.620(1)	11.621(1)	11.625(1)	11.639(1)	11.663(1)
U	1.07(2)	0.95(5)	0.94(6)	0.96(5)	1.00(6)	1.03(6)
V	−0.80(6)	−0.76(5)	−0.75(5)	−0.75(5)	−0.77(5)	−0.78(5)
W	0.29(1)	0.28(1)	0.28(1)	0.28(1)	0.28(1)	0.28(1)
$Zero$	−0.057(3)	−0.058(3)	−0.057(3)	−0.056(3)	−0.055(3)	−0.057(3)
R_{prof}	2.00	1.61	1.75	1.60	1.65	1.60
R_{Bragg}	5.18	5.11	5.56	5.25	6.23	6.73

scheme used may result in a different statistical bias (see Wilson 1976). The data used for Rietveld refinement are a set of step-scan observations y_i, sitting on the background B_i. Each observation is assigned a statistical weight

$$w_i = 1 \, / \, \sigma^2(y_i) + \sigma^2(B_i) \tag{3.44}$$

where σ^2 is the variance of the appropriate quantity. Estimated standard deviations (e.s.d.) in Rietveld refinements are in general quoted better than those from integrated intensity refinements. It has been pointed out (Cooper 1983), that the quoted uncertainties in Rietveld refinements may be unreliable because of systematic errors. The reasons are not well understood. Pawley (1980) suggests that the e.s.d. calculated by the Rietveld method should be multiplied by a factor in the order of 2 to 3. Since B_i is obtained by graphical means its variance is not known and is generally arbitrarily set to zero. From this it follows that data, specifically e.s.d. derived by the Rietveld method are only meaningful if the values y_i used for the analysis are derived by subtracting background values B_i beforehand.

3.7.5
Conclusion

In conclusion we find that both method lead to the same structural results, if applied properly. On the other hand systematic differences are easily introduced into the analysis of Rietveld refinements so that the values obtained for structural parameters will not be exactly the same as those obtained from integrated intensity of the same data (Sakata and Cooper 1979). Those differences in these values may not be statistically significant for a fairly simple diffraction pattern, and it is possible that they may be larger for more complex patterns. Also the standard deviations σ are determined incorrectly in the profile refinement method. In most examples which have been considered by Sakata and Cooper (1979) they have been underestimated by a factor of at least two. Especially the R-values are far too low and far from reality.

The situation is even more complicated when peaks overlap. In addition, it should be noted that the calculation, i.e. refinement assumes that he intrinsic diffraction profile, i.e. without line broadening by the sample properties, is independent of the structural parameters. If the intrinsic line broadening is significant, for example by small particle size or internal strain, the possible variation of the structure factor with angle needs to be considered.

The structural parameters (x, y, z, B) influence only the values of $I(hkl)$ and it is normally only these which are of direct interest. Thus for a single resolved peak each intensity measurement on the profile gives an estimate of the same Bragg intensity and gives different information concerning the profile parameters alone. Hence no additional structural information can be gained by increasing the number of points on the profile, other than which can be gained by an effective increase in the precision of the determination of the Bragg intensity.

In *ab initio* structure determinations by "direct methods" powder diffraction reaches a point where it is fully competitive with single crystal data. This means that an unknown structure shall be solved *ab initio* from the diffraction data, and this does require separated intensities for every reflection. This can be achieved only with the (first stage) in the Two Stage method. Then the common techniques like Patterson calculations, heavy atom method, etc. and also direct methods are used (see for example Giacovazzo 1996).

3.8
Analysis of Quartz

Quartz is especially well suited to demonstrate and test the Two Stage method and to find the best experimental conditions for getting the best results. At the beginning of this project data were collected in the laboratory using sealed X-ray tubes with Cu Kα radiation, later it was extended to synchrotron X-rays. For the crystal structure refinement the program POWLS was used. Before any crystallographic problem can be approached three steps of equal importance must be dealt with:

- sample preparation
- experimental setup and experimental conditions
- separation of peaks by computer assisted profile fitting

It must be emphasized that each step before the next one is always a crucial one. No matter how much care is placed on the experiment, it cannot compensate for any negligence made in the specimen preparation. And needless to say, no computer program, no matter how sophisticated, can correct for failures made in specimen preparation and in the experiment.

3.8.1
Sample Preparation

It is evident that the powder specimen preparation is a crucial factor in achieving good results. Random orientation of crystallites is essential to determine truely relative intensities. Complete randomness is hardly to achieve and corrections are necessary in the experiment as well as in the analysis. In the generally used Bragg-Brenanto geometry with flat specimens only about $0.1\ cm^3$ volume takes part in the diffraction and the surface layer is the main contributor. It is different in neutron diffraction where virtually the whole, generally cylindrical volume contributes, making it more likely to achieve an acceptable random distribution.

It requires a great deal of experience and patience to prepare specimens of quality suitable for structure determination if it is meant to compete with analysis from single crystals. A major problem arises from larger particles in the specimen and micro-absorption. To minimize this problem grinding and sifting is essential. The best results were obtained with particles 5 µm and less, which can be obtained for example with micromesh sifters. In our experiments they were made of 0.035 mm thick nickel foils with 5, 10, and 20 µm square holes, bought from Bucknee-Mears, Minneapolis, Minn., USA. In a systemic study on the influence of grain size on the results particle sizes >30 µm, 20–30 µm, 10–20 µm and <5 µm were selected and investigated. As a sample holder a single crystal silicon plate cut parallel to (510) was used and is recommended. Such plates are available commercially. This orientation gives no reflection and gives virtually no background. The samples of quartz used for these experiments were prepared from synthetic high grade oscillator plate quality single crystals.

3.8.2
Experiments

The experiments were done on a Philips horizontal diffractometer in the laboratory with a graphite monochromator, fine Soller slits and evacuated beam paths. In synchrotron experiments a channel cut monochromator was used. During the experiments the specimens were rotated around the normal to the specimen plane. Synchrotron X-rays offer several advantages of over conventional X-ray tube focusing methods:

- Any wavelength in the range from about 0.5 to 2.5 Å can easily be selected to maximize peak-to-background, or to choose a specific wavelength for anomalous dispersion measurements
- A single wavelength produces single profile reflections instead of $K\alpha$ doublets. A good monochromator gives a very small wavelength spread. The parallel beam collimator eliminates a number of geometrical aberrations making $W \cdot G$ simpler
- The reflections are virtually symmetrical Gaussian type profiles whose shape is constant over a wide 2Θ range. This greatly simplifies the profile fitting procedure and the determination of specimen contribution to profile broadening caused by small particle size, strain, stacking faults, etc., for Warren-type profile analysis
- As a further advantage high precision determination of lattice parameters is easily possible
- The primary beam intensity is approximately uniform over a wide range of wavelength and is orders of magnitude higher than from X-ray tubes
- Using a short wavelength the reciprocal space can be explored far out, for example with 1 Å to $\sin\Theta / \lambda = 0.99$ Å^{-1}. The limit with Cu $K\alpha$ is 0.65 Å^{-1}

Figure 3.35 shows a section of a typical quartz pattern measured with 1 Å synchrotron X-rays. The diffractometer was using 2:1 step-scanning with a step size $\Delta(2\Theta) = 0.02°$. The *FWHM* was 0.17°. The experiment covered the range $d = 1.37$ to 0.54 Å, within $2\Theta = 12$ to 135°, corresponding to $\sin\Theta / \lambda = 0.105$ to 0.942 Å^{-1}. Using Cu $K\alpha$ the limit is $d = 0.79$ Å giving only 65 reflections instead of 161 resolved reflections recorded here with intensities determined by profile fitting. The insert depicts a profile fitted detailed section of one peak cluster around $2\Theta = 116°$. The differences between observed and profile fitted intensities are shown at half height.

The instrumentation in synchrotron experiments is similar to conventional diffractometry with a few important exceptions:

- A channel cut or double crystal monochromator (germanium or silicon) is used to select the wavelength form the white beam source. It has the important property that the direction and position of the monochromatic beam remains fixed making it unnecessary to re-align the instrument or re-calibrate over a wide range of wavelengths
- The diffracted beam can be defined by a crystal analyzer, receiving slit or horizontal parallel slits
- The synchrotron X-ray beam is highly polarized in the horizontal plan, a vertical scanning diffractometer avoids this problem

3.8.3
Sample Rotation during Experiments

It is hardly possible to obtain a specimen with fully random orientation of crystallites as required by theory. Some improvement can be obtained by rotating the sample around the horizontal plane, e.g. around the scattering vector. A second correction is computational. In most least squares programs this correction is labeled – wrongly –

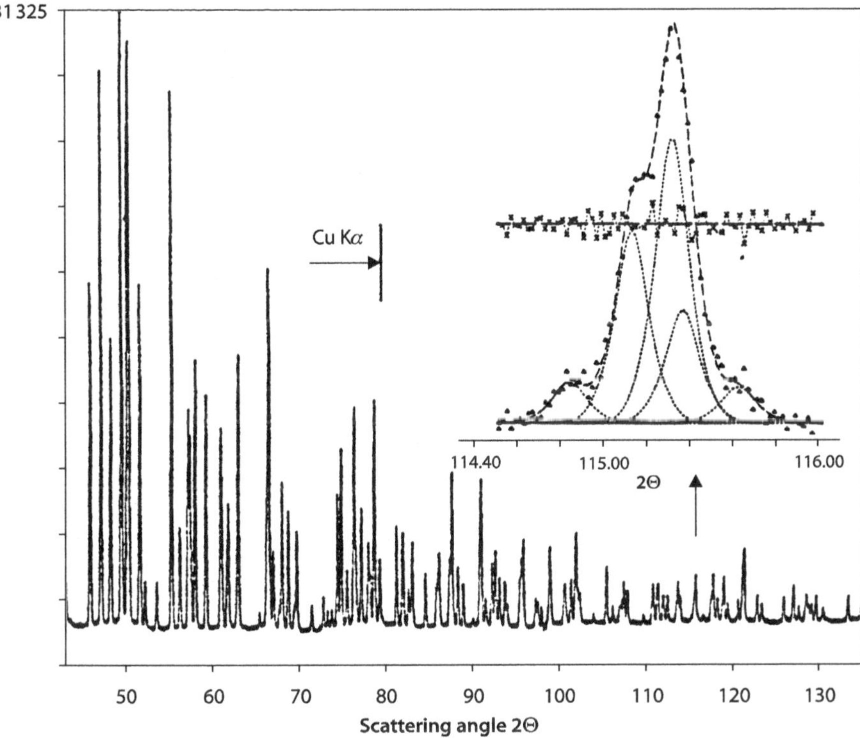

Fig. 3.35. Diffraction pattern of quartz measured with 1 Å synchrotron X-rays. The *insert* gives a profile fitted section around 2Θ = 116°

"preferred orientation", In reality it is a correction for the "non-random" particle distribution. The influence of the particle size on the quality of the diffraction pattern is demonstrated in Fig. 3.36 with two samples of silicon <5 μm and 10–20 μm. The specimens were rotated first rapidly, about 60 r min^{-1} (right side), and then, left side, with a reduced rotational speed of 1/7 r min^{-1}. With the larger fraction of 10–20 μm the contribution of individual grains is clearly visible. The difference is as large as 13.5%, versus only 1% with small crystallites <5 μm.

3.8.1
Crystal Structure Refinement with POWLS of α-Quartz Collected with Cu Kα Radiation

The measurements done in the laboratory served to test the power of the Two Stage method in comparison with results from single crystal analysis and also to explore the power of laboratory experiments, still the most used experimental facilities.

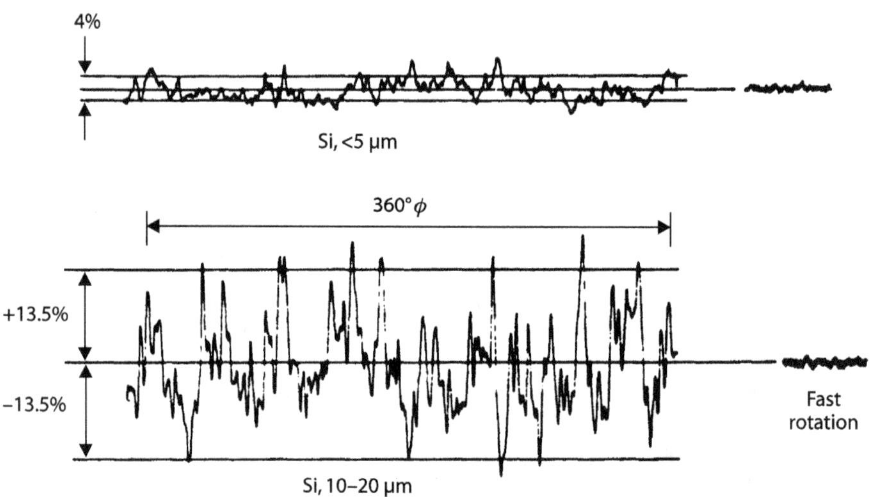

Fig. 3.36. Variation of peak intensity of silicon powder (111), Cu Kα, with azimuthal rotation

As a first example several sets of α-quartz with particle sizes <5 μm, 10–20 μm and >30 μm were measured with Cu Kα and refined with POWLS. The data set of the <5 μm sample is discussed further. It was measured over a range 2Θ = 19 to 109° leading to 40 measurable reflections, which could be resolved, containing 62 (hkl) Miller planes. The symmetry of quartz causes intrinsic superposition of planes with $d(hol) = d(ohl)$ but with different intensities $F(hol) \neq F(ohl)$. The subroutine FUNNY together with OLAPP in POWLS is programmed to handle such cases. Table 3.14 gives observed and calculated intensities. The wide range of measured integrated intensities, ranging from I_{max} = 141 102 counts to I_{min} = 28 counts, provided a good test of the profile fitting method in deriving accurate intensities. The structure analysis was done by refining four positional parameters: x(Si), x(O), y(O), and z(O), the scale factor and two (for isotropic) or ten (for anisotropic) values for the temperature factors with the results listed in Table 3.15 in comparison with results from single crystal data in the literature (Levien et al. 1980; LePage and Donnay 1976; Smith and Alexander 1963). The R_{Bragg}-value was 0.84%. With the data sets containing larger particles the R-values were much higher, up to 3%. The agreement of parameters is good, but the e.s.d. are considerably higher. Although there is good agreement of the β-values with single crystal data, temperature parameters from powder diffraction data are not well suited for deriving either isotropic or anisotropic temperature factors. This holds equally well for Rietveld refinements. Many systematic errors in the experimental method and sample preparation can be generally accounted for by an exponential function and such errors are consequently very likely to be absorbed in the temperature factors.

In a similar experiment corundum was studied. The sample was Linde A synthetic α-Al$_2$O$_3$, <5 μm particles. The refinement was made with 24 front reflections, up to 2Θ = 94°. Observed and calculated values are given in Table 3.16. The intensities range from I_{max} = 36 531 to I_{min} = 115 counts. A correction for "preferred orientation" was included in the refinement. The final R_{Bragg}-value was 1.4%.

Table 3.14. Comparison of observed and calculated intensities of α-quartz, α-SiO$_2$ (Cu Kα)

hkl	I(obs)	I(calc)	hkl	I(obs)	I(calc)
100	29 315	29 307	220	2 229	2 272
101,011	141 102	141 114	213,123	4 282	4 203
110	10 338	10 343	221	954	1 079
102,012	9 267	9 250	114	3 350	3 458
111	4 654	4 565	310	3 885	4 027
200	6 992	7 090	311,131	2 729	2 876
201,021	4 860	4 904	204,024	373	431
112	18 297	18 131	222	28	81
003	345	317	303,033	358	495
202,022	5 567	5 403	312,132	4 506	4 335
103,013	2 122	2 151	400	831	801
210	314	373	105,015	2 049	2 133
211,121	13 133	13 164	401,041	1 506	1 549
113	2 262	2 415	214,124	2 340	2 435
300	590	692	223	2 098	2 144
212,122	8 233	8 068	402,042	286	258
203,023	9 465	9 568	115	1 527	1 471
301,031	5 683	5 822	313,133	1 044	722
104,014	2 831	2 868	304,034	570	407
302,032	4 090	3 993	320	1 139	1 092

3.8.5
Refinement of α-Quartz Data collected with Synchrotron X-Rays

Synchrotron radiation offers the possibility to explore the influence of large data sets reaching far into reciprocal space. Improvement of R-values and standard deviations were expected when using more observations. For this purpose data were collected with $\lambda = 1.0020$ Å up to $2\Theta = 135°$, $s_{max} = 0.92$ Å^{-1}, with $s = \sin\Theta / \lambda$, and in a second experiment with $\lambda = 1.28284$ Å, to improve the spatial resolution.

The data set with $\lambda = 1.0020$ Å was collected with a step width $\Delta(2\Theta) = 0.02°$ in the angular range $2\Theta = 45$ to $135°$. Because of the polarization of the synchrotron beam in the horizontal plane measurement were performed in a vertical scanning mode going from high angles to low angles to avoid any possible slipping in the gears carrying the counter. This experiment had to be terminated at $2\Theta = 45°$ because of a failure in the synchrotron accelerator. The diffraction pattern is shown in Fig. 3.37. Note the change in the intensity scale. Even at high angles, well beyond the Cu Kα sphere, ending here at $2\Theta = 82°$, the peaks are well resolved. Even at high angles they are well above background with good intensities. The halfwidth is $0.17°$. The peaks were analyzed and separated first with PROFAN in sections and then with FULFIT for the whole diagram.

Table 3.15. Comparison of structural parameters of quartz, α-SiO$_2$ (Cu Kα). Intensities from Table 3.14

	This work			Levien et al. (1980)[a]	Le Page and Donnay (1976)[a]	Smith and Alexander (1963)[a]
	Isotropic temperature factors	Anisotropic temperature factors				
x(Si)	0.4713(5)	0.4724(8)	0.4723(8)	0.4697(1)	0.46987(9)	0.4698(3)
x(O)	0.4144(3)	0.4150(10)	0.4151(9)	0.4135(3)	0.4141(2)	0.4145(8)
y(O)	0.2660(6)	0.2657(9)	0.2657(9)	0.2669(2)	0.2681(2)	0.2662(7)
z(O)	0.1205(5)	0.1198(6)	0.1198(6)	0.1191(2)	0.1188(1)	0.1189(4)
B(Si)[b]	0.96(5)			0.62(2)		
B(O)	0.86(9)			1.05(2)		
β_{11}(Si)[c]		1.69(20)	1.66(20)	0.93(2)	0.66(1)	0.48(6)
β_{22}(Si)		0.98(37)	0.96(33)	0.78(2)	0.51(2)	0.27(7)
β_{33}(Si)		0.28(21)	0.29(15)	0.49(2)	0.60(1)	0.63(6)
β_{13}(Si)		−0.05(8)	−0.02(8)	−0.001(7)	−0.03(1)	0.04(4)
β_{11}(O)		1.70(54)	1.56(47)	1.90(6)	1.56(4)	1.28(16)
β_{22}(O)		0.80(26)	0.84(26)	1.44(5)	1.15(3)	1.05(14)
β_{33}(O)		0.30(44)	0.58(39)	0.83(3)	1.19(3)	1.28(12)
β_{12}(O)		0.83(37)	0.76(35)	1.06(5)	0.92(3)	0.69(12)
β_{13}(O)		0.15(23)	0.07(21)	−0.25(3)	−0.29(3)	−0.35(12)
β_{23}(O)		0.70(20)	0.50(19)	−0.35(3)	0.46(2)	−0.44(10)
R(Bragg) (%)	1.00	0.84	1.13	1.9	1.57	3.3
Number of reflections	47	47	62	210	342	112
Number of observations	31	31	40			

[a] Results from single-crystal methods.
[b] Unit of $B = \text{Å}^2$.
[c] All anisotropic temperature factors β are given $\times 10^2$.

The profile R-values are typically 1–2%. With this very short wavelength 280 peaks could be obtained, 157 of them single peaks. Due to the trigonal symmetry of quartz with $I(hkl) \neq I(khl)$ the remaining peaks came from intrinsically overlapping Miller planes at the same position but with different intensities. At high angles many of the peaks had quite low intensities and consequently poor counting statistics, but they are clearly visible in the pattern and well above background.

In a second synchrotron duty cycle a second data set was collected at a longer wavelength $\lambda = 1.28284$ Å and shorter step width $\Delta(2\Theta) = 0.01°$ to increase the spatial resolution. A new and improved narrower collimator system gave increased spatial resolution with a profile $FWHM$ now 0.05° against 0.17° before. The angular range $2\Theta = 15$ to 115° ($s_{\max} = 0.66$ Å^{-1}) was covered. Exposure time was $t = 1$ s per step. Particle size

Table 3.16. Comparison of observed and calculated intensities for corundum, α-Al$_2$O$_3$ (Cu Kα). Shown are also the structural parameters in comparison with literature data

hkl	I(obs)	I(calc)	hkl	I(obs)	I(calc)
012	21 535	21 505	300	19 920	19 989
104	33 064	33 040	125	504	412
110	14 082	14 062	208	610	571
006	236	288	1.0.10	6 468	6 133
113	36 351	36 424	119	2 909	2 904
202	400	482	217,220	2 444	2 597
024	17 936	17 896	036,306	310	332
116	34 389	34 601	223	1 860	1 664
211	860	862	131	115	92
122	1 210	1 142	312	1 650	1 301
018	2 400	2 831	128	1 000	1 028
214	13 700	13 728	0.2.10	3 299	2 733

	This study	Will et al. (1982)	Cox et al. (1980)
x(O)	0.3057(5)	0.3065(5)	0.3060
z(Al)	0.3523(1)	0.3526(3)	0.3520
B(O)[a]	1.09(7)	0.06	0.15
B(Al)	0.75(8)	0.28	0.22

[a] Unit of $B = \text{Å}^2$.

was 5–10 μm. Ten thousand data points were recorded in about 3 hours. The intensities had a range from 36 866 to 8 counts, a ratio of about 4 600:1. The lattice parameters were determined by least-squares refinement to be $a = 4.91239(4)$, $c = 5.40385(7)$ Å. In the pattern 72 peaks from 119 Miller planes (due to the intrinsic superposition of planes with $d(h0l) = d(0hl)$ but with different intensities $I(h0l) \neq I(0hl)$) could be separated by profile fitting. The individual profile R-values obtained with PROFAN are in the range 1.1 to 2.0%.

Refinement was done with the program POWLS using the profile fitted intensities. Due to the highly polarized synchrotron beam no polarization correction is necessary. Correction for "preferred orientation" was included in the refinement, since despite careful sample preparation it is not possible to arrive at a perfect random distribution of the crystallites. By checking several possible planes of orientation (211) gave the best results and was finally used.

In order to test the power of the method compared with single crystal analysis and also the limits of synchrotron experimental techniques several refinements with POWLS were carried out varying the size and range of the data sets. The refinement

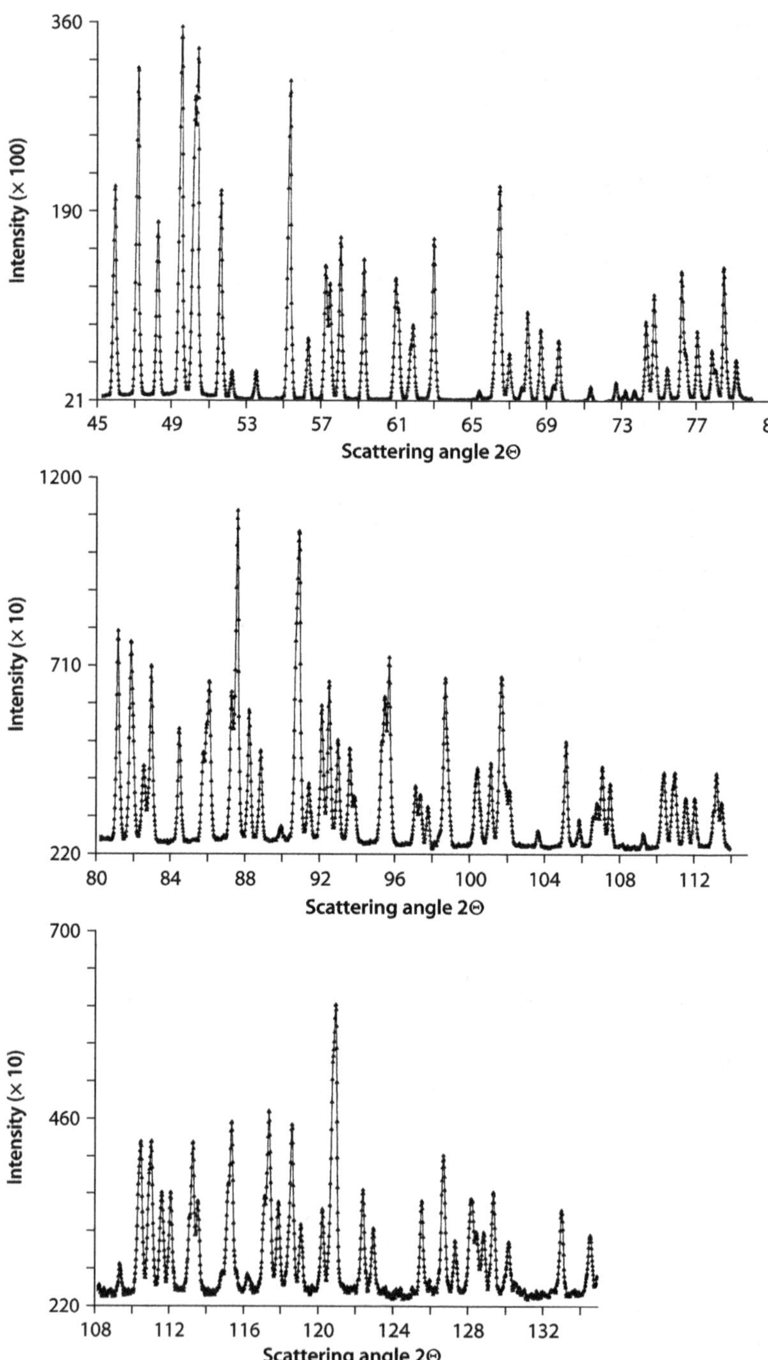

Fig. 3.37. Diffraction pattern of quartz measured with 1.002 Å synchrotron X-rays. Range in 2Θ is 45 to 135°. Note the change in the intensity scale

was begun with a data set containing only the first 30 peaks up to $2\Theta = 75°$ ($s_{max} = 0.47$ Å^{-1}), still beyond the range with Cu Kα radiation). Refinement was then continued using larger data sets as listed in Table 3.17: 40, 50, 60 and 72 reflections containing 46, 81, 98 and 119 Miller planes. The increased number of data did not significantly change the parameters or the standard deviations, and this was surprising. Especially the R_{Bragg}-values did not improve, on the contrary they became slightly larger. This may be due to the inclusion of poorer data as more weak peaks are included at higher angles. Even with the short data set with 46 hkl planes the system is fivefold overdetermined and the results are therefore well justified. The agreement with published data (Levien et al. 1980) is excellent. The parameters are listed in Table 3.18. It is interesting to note that the inclusion of data further out in reciprocal space, i.e. including more observations in the least squares calculations has no or only little influence on the parameters and not even on the e.s.d. It seems therefore that intensities up to $2\Theta = 90°$, equivalent to $s = 0.7$ Å^{-1}, are sufficient to reach good and reliable results. The theoretical full range of the Cu-sphere is only $s = 0.65$ Å^{-1}.

Also here the refinement calculation with POWLS was begun with a smaller, lower angle data set up to $2\Theta = 90°$ with the first 60 peaks coming from 150 hkl planes. Interesting the best results were obtained with 40 observations from 64 hkl planes, which resulted in $R = 1.41\%$. With 8 variables the system is already 5-fold overdetermined, therefore the limitation of the data set is well justified. With the full data set up to $2\Theta = 115°$ we have 72 observations and 119 hkl planes and the system is 15-fold overdetermined. The resulting R-value is $R = 1.68\%$.

For the Debye-Waller factors on the other hand reflections out at small d-values, i.e. high scattering angles improve the results. This is especially true, if we remember that bonding effects influence the data at low diffraction angles and are negligible at higher angles. This is known and used in deformation density studies, where the data set are split into high order (HO) and low order (LO) reflections.

Table 3.17. Effect of number of reflections on structure refinement of quartz

Number of reflections	30	40	50	60	72
Number of planes	46	46	81	98	119
$\sin\Theta / \lambda_{max}$ (Å^{-1})	0.464	0.520	0.589	0.612	0.656
R_{Bragg} (%)	1.35	1.60	1.90	2.08	2.20
x(Si)	0.4703(6)	0.4698(5)	0.4699(5)	0.4699(5)	0.4701(4)
y(Si)	0	0	0	0	0
z(Si)	0.6667	0.6667	0.6667	0.6667	0.6667
x(O)	0.4140(8)	0.4139(7)	0.4139(7)	0.4138(7)	0.4139(7)
y(O)	0.2677(8)	0.2674(8)	0.2676(8)	0.2675(7)	0.2674(7)
z(O)	0.7855(7)	0.7856(7)	0.7858(7)	0.785886)	0.7856(6)
B(Si) (Å^2)	0.47(7)	0.39(4)	0.31(3)	0.30(3)	0.29(2)
B(O) (Å^2)	0.67(13)	0.87(9)	0.89(9)	0.87(8)	0.85(6)

Table 3.18. Comparison of structural parameters derived from powder and single crystal diffraction data

	Powder isotropic thermal factors[a]	Powder anisotropic thermal factors[a]	Single crystal[b]
Number of observations	72	72	663
Number of planes	119	119	210
R_{Bragg} (%)	2.20	1.60	1.60
x(Si)	0.4701(4)	0.4704(4)	0.4697(1)
y(Si)	0	0	0
z(Si)	0.6667	0.6667	0.6667
x(O)	0.4139(7)	0.4136(6)	0.4135(3)
y(O)	0.2674(7)	0.2676(6)	0.2669(2)
z(O)	0.7856(6)	0.7857(5)	0.7857(2)
β_{11}(Si)		0.59(7)	0.93(2)
β_{22}(Si)		0.3(1)	0.78(2)
β_{33}(Si)		0.43(8)	0.49(2)
β_{12}(Si)		0.17(5)	0.39(1)
β_{23}(Si)		−0.2(1)	−0.00(1)
β_{13}(Si)		−0.10(5)	−0.001(7)
B(Si) (Å^2)	0.29(2)		0.62(2)
β_{11}(O)		0.8(2)	1.90(6)
β_{22}(O)		0.9(3)	1.44(5)
β_{33}(O)		0.9(2)	0.83(3)
β_{12}(O)		1.2(3)	1.06(5)
β_{23}(O)		−0.2(3)	−0.35(3)
β_{13}(O)		0.0(2)	−0.25(3)
B(O) (Å^2)	0.85(6)		1.05(2)
GP	0.21(2)		

[a] Units are $\beta_{ij} \times 10$.
[b] Levien et al. (1980).

In conclusion we find, that observed and calculated intensities derived by FULFIT, the profile fitting program gives intensities not influenced by a model structure. They are therefore real observed intensities, directly comparable to single crystal data. And only such intensities can be used for further Fourier calculations or for *ab initio* structure determinations (Giacovazzo 1996). Also the R_{Bragg}-values are "real" R-values.

3.9
Texture Analysis with the Two Stage Method Using Neutron Diffraction

3.9.1
Introduction

Texture analysis is an important tool in science as well as in industry, where it is routinely used with X-rays for example in metal manufacturing. There are limitations when using X-rays: first because of the high absorption of X-rays by matter, and second because texture measurements require rotation of the specimen in space through all Eulerian angles. In conventional measurements with X-rays the (generally used) single channel detector is placed at one specific peak position and the (maximum) intensity is determined as a function of the three Eulerian angles ϕ, χ, and ω. Those data will then be processed to calculate pole figures and ODFs, orientation distribution (functions). Metals, crystallizing with high symmetry and small unit cells are the samples where texture analysis has been used for many years, and very successful. Those materials have well resolved reflections with no or rarely overlap of peaks and the detector records the intensity from one *hkl* plane. Such measurements pose no serious problems. The situation is different with low symmetry crystals and/or large unit cell parameters, which is typical for minerals and rocks, i.e. in geosciences. As an additional problem these samples are composed quite common of several minerals and high peak overlap is a common feature. Since texture analysis is an important tool of research in geosciences it asked for solutions.

Texture analysis based on X-ray diffraction is not really suited for investigations in geosciences. Neutron diffraction on the other hand is a powerful means and opens a new way to study textures also of rocks, or generally of materials with low symmetry or large unit cells with overlapping reflections, and also, very important, of multiphase samples. A comparison between the two radiations in their use for texture measurements has been published by Engler et al. (1993). A number of factors are in favor of neutron diffraction:

- Large samples, about 1 cm in diameter can be measured without difficulties due to the low absorption of neutrons by the specimen. Corrections for absorption are negligible
- The investigations are non-destructive because it does not require cutting or polishing the sample
- Due to the high penetration of neutrons the texture of the whole specimen is investigated in transmission: *global texture*, which is in contrast to X-ray measurements, where only *local textures* on the surface of the specimens can be determined
- Nuclear scattering is independent of 2Θ and thus allows one to measure pole figures even on high index reflections

Despite these obvious advantages and potentials texture measurements based on neutron diffraction was not seriously used in the past. There were sporadic attempts by Brockhouse in 1953, and later also by some other authors, but without broader acceptance. For more information see Wenk et al. (1984). A systematic and effective use of neutron diffraction for texture analysis required however extensive methodical and instrumental developments with improvements in hardware as well as software. Such are:

- the development of position sensitive detectors to overcome long measuring times due to the limited neutron flux
- methods for data collection adapted to the specific needs in neutron diffraction experiments
- the development of programs for profile fitting and peak separation to open the way for calculating pole figures followed by the calculation of orientation distributions of the crystallites

Without position sensitive detectors texture analysis with neutrons can never reach a wider clientele because of unrealistic long times required for the measurement of one specimen. Position sensitive neutron detectors were therefore the most urgent requirements and was the first step in opening this line of research. Such developments were done in the KFA Jülich by the Neutron Diffraction Group of the Mineralogical Institute of the University Bonn (Höfler et al. 1985; Jansen E. et al. 1985, 1988; Schäfer et al. 1988, 1991, 1992a; Will et al. 1986a,b, 1988a,b, 1989, 1990c, 1992a; Wenk et al. 1984). The technique and examples presented here are therefore limited to results obtained by this group. The full advantages of texture analysis by neutron diffraction with a position sensitive detector are:

- The whole pattern, or at least a large section, is recorded, compensating for the intensity disadvantage against X-rays
- Several pole figures, even of different compounds, for example different minerals in a multiphase specimen are measured simultaneously reducing the time required for pole figures. Experience has shown that up to 28 pole figures from up to 3 minerals could be measured simultaneously
- The full reflection profile is recorded, not just the peak maximum, thus avoiding errors arising when peaks are too close to each other and overlap each other seriously
- The data are collected in transmission geometry giving the complete pole figures, whereas in X-ray diffraction pole figures are either incomplete or have to be composed of transmission and reflection data in two separate measurements
- Using profile fitting methods allow the separation of reflections too close to be measured individually. They can be separated and then used for individual pole figure calculations
- Low symmetry samples can be studied
- Multiphase samples can be studied and the pole figures, and more important the ODFs of the different components determined
- The time spent for the investigation of one specimen is drastically reduced to the order of hours

3.9.2
The Position Sensitive Detector JULIOS

The detector JULIOS is a linear position sensitive detector developed at the KFA Jülich and published in a number of papers (see for example Will et al. 1989, 1994). JULIOS is a solid state scintillation detector based on the nuclear reaction

$$\,^1_0\mathrm{n} + \,^6_3\mathrm{Li} = \,^3_1\mathrm{t} + \,^4_2\alpha + 4.6\,\mathrm{MeV} \tag{3.45}$$

(t for tritium). Ce atoms imbedded in the ^{6}Li-glas are excited by the α-particles giving a light flush which is transmitted by a light coupler to a row of 24 photo multipliers. From the different light intensities reaching three of the 24 photo multipliers the position of the event is calculated. The detector is 682 mm long. At a (typical) distance of 100 cm from the specimen the 2Θ range is $36°$. Figure 3.38 shows a photograph of the JULIOS detector.

3.9.3
Data Collection

Neutron fluxes are in general moderate with about $2 \times 10^{14}\,\mathrm{n\,cm^{-2}\,s^{-1}}$ in the core at the DIDO research reactor at Jülich. With a Cu single crystal monochromator using

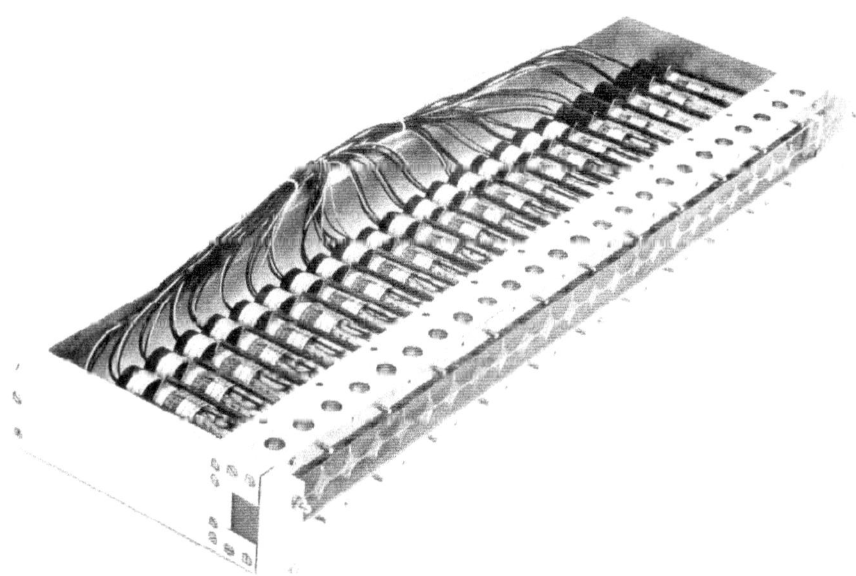

Fig. 3.38. Photograph of the PSD JULIOS

the (111) plane and a wavelength of $\lambda = 1.293$ Å the flux at the specimen was about 10^6 n cm^{-2} s^{-1}. The beam size at the specimen position was 20 mm × 40 mm. The size of the specimen is determined by the size of the primary beam. The sample size is typically 1 cm × 1 cm × 1 cm. The experimental setup at Jülich is shown schematically in Fig. 3.39.

Data are required for the whole hemisphere. Therefore the sample must be rotated in Eulerian space around the three Eulerian angles ϕ, χ, and ω. (Fig. 3.40) Spherical, or nearly spherical specimens are preferred. Because of the negligible absorption of neutrons cubes are equally well suited. Measurements with neutrons are always made in transmission in a step scanning mode on concentric circles with increments in the azimuth $\Delta\phi = 5°$ for the outer parts, $0° < \chi < 25°$, $\Delta\phi = 10°$ for $30° < \chi < 55°$ and $\Delta\phi = 20°$ for $60° < \chi < 90°$. Increments in pole distance $\Delta\chi$ are 5°. This gives a total of 757 data points, i.e. steps on the hemisphere. The scanning schedule is based on "equal area scanning". The position sensitive detector is kept constant at the specific Bragg angle of a selected group of *hkl* reflections.

For one specimen typically a total of about 800 single patterns covering a range in 2Θ of 24° are recorded containing (in geological samples) typically between 9 and 30 peaks. Typical exposure times are minutes for one pattern and this means about 8 to 24 hours for one specimen (with 9 to 30 peaks, i.e. pole figures). The individual patterns collected in a short time and with a moderate neutron flux are naturally of "poor qual-

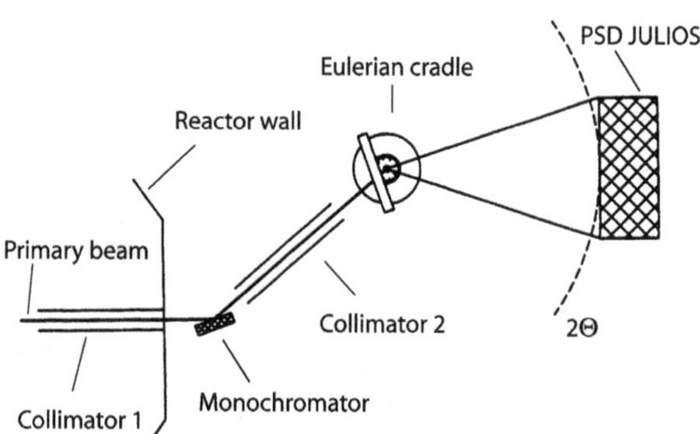

Fig. 3.39. Schematic setup of the texture goniometer

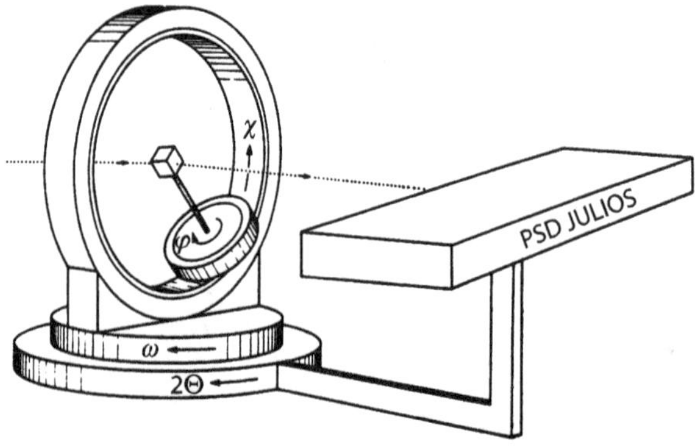

Fig. 3.40. Schematics of the four-circle diffractometer with the PSD JULIOS

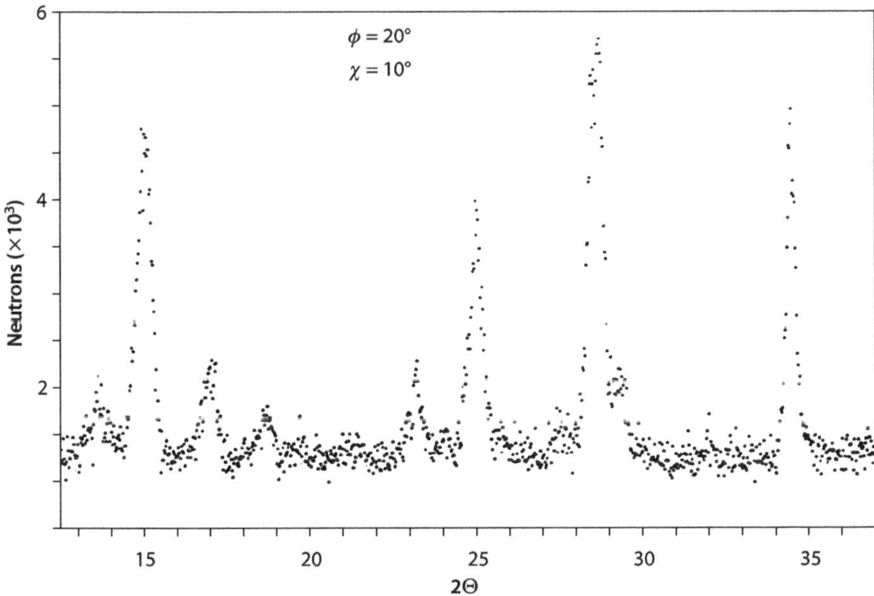

Fig. 3.41. Individual, single position diffraction pattern of a hematite sample: $\phi = 20°$, $\chi = 10°$. Time spent was 3 minutes

ity" in the common sense. Figure 3.41 shows such a single position pattern of a hematite sample covering $2\Theta = 12$ to $37°$ collected at $\phi = 20°$ and $\chi = 10°$. The *FWHM* is 0.6°.

3.9.4
Data Evaluation

A typical data set consists of about 800 single diffraction patterns. The evaluation of one such complete set takes between 30 and 100 minutes using a 386 processor with 20 MHz frequency. The data evaluation is done interactively in front of a screen using a personal computer in a number of steps with the program package PROFAN-PC (Merz et al. 1990):

- Adding up all individual patterns to a so-called sum diagram. Figure 3.42 shows the hematite specimen from Fig. 3.41 as an example. The sum pattern is naturally of good quality and can be analyzed by profile fitting procedures. Figure 3.43 gives a 3-dimensional representation of a set of individual diffraction diagrams of the hematite specimen at $\phi = 100°$, $\chi = 0°$ to 180°
- Determining and subtracting the background, either by orthogonal polynomials according to Steenstrup (1981) or with a cubic spline. Figure 3.44 gives as an example a pattern fitted with a polynomial of degree nine. The sample was an alloy, ZnAl
- Standard profile analysis and fitting, again interactively. In a first step the positions of expected peaks are marked by cross hairs (see Fig. 3.45a), and then fitted by least squares procedures. (Fig. 3.45b). The results are peak positions, the profile function best suited to describe the peaks, and the *FWHM*

- In the next step the operator goes back to the individual diffraction patterns to calculate with the known parameters from the sum-pattern: profile function, 2Θ positions, *FWHM*, the intensities of the individual patterns, i.e. intensities as a function of ϕ, χ, and ω for each reflection
- These data are then transferred into the pole figure program to calculate pole figures. Figure 3.46 gives two such pole figures for hematite
- If required the pole figures can be used to calculate the orientation distribution of the crystallites (ODF)

3.9.5
Examples of Texture Analysis

3.9.5.1
Rolled Titanium Steel Sheets

Texture analysis of rolled metal sheets is the main application with X-rays. To study the possibility in using neutrons titanium sheets of different rolling degrees: 40, 50 and 80% were studied. The sheets were only about 1 mm thick, and therefore for intensity reasons cubical specimen had been prepared by stacking 10 sheets of 10×10 mm^2 cross section on top of each other. Intensities have been collected for pole figure calculations of the most relevant reflections (hexagonal symmetry): 100, 002, 101 and 110. The influence of rolling degree can be seen best by comparing the 002 pole figures, as shown in Fig. 3.47. Depending on the degree of rolling

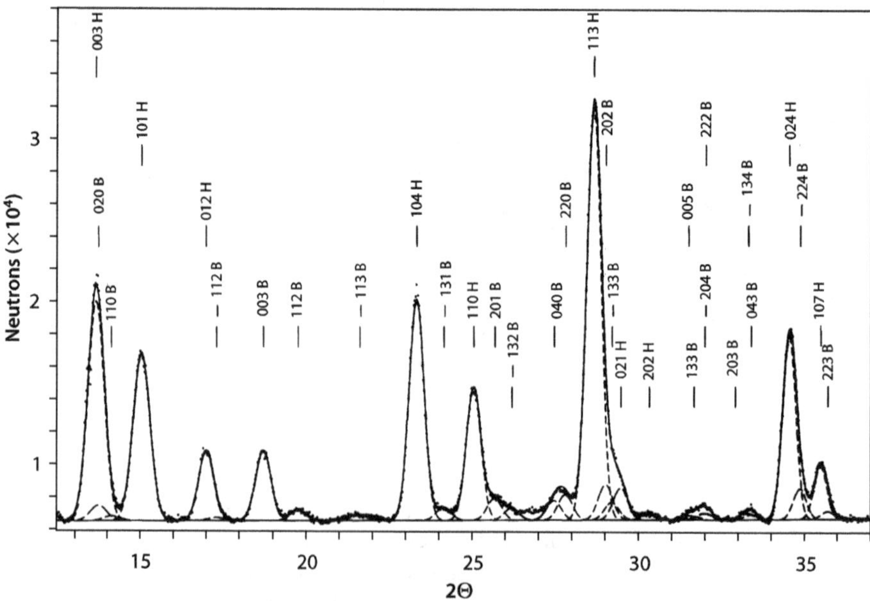

Fig. 3.42. Typical "sum" pattern generated by adding up all 621 individual patterns like that in Fig. 3.41. The specimen had two phases, H = hematite, B = biotite. Background has been subtracted

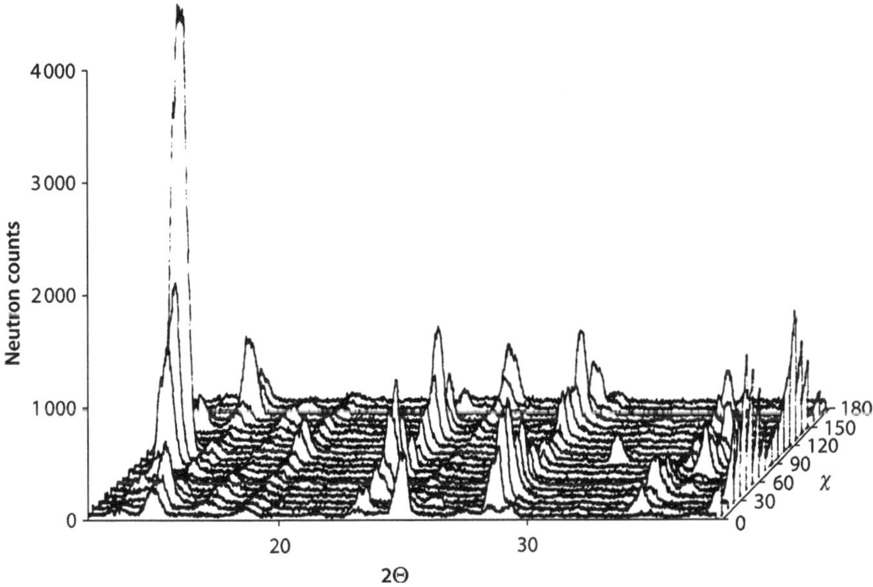

Fig. 3.43. A 3-dimensional representation of a set of individual diffraction diagrams of a hematite speci-men at $\phi = 100°$, $\chi = 0°$ to $180°$. The variation with rotating the specimen is dramatic

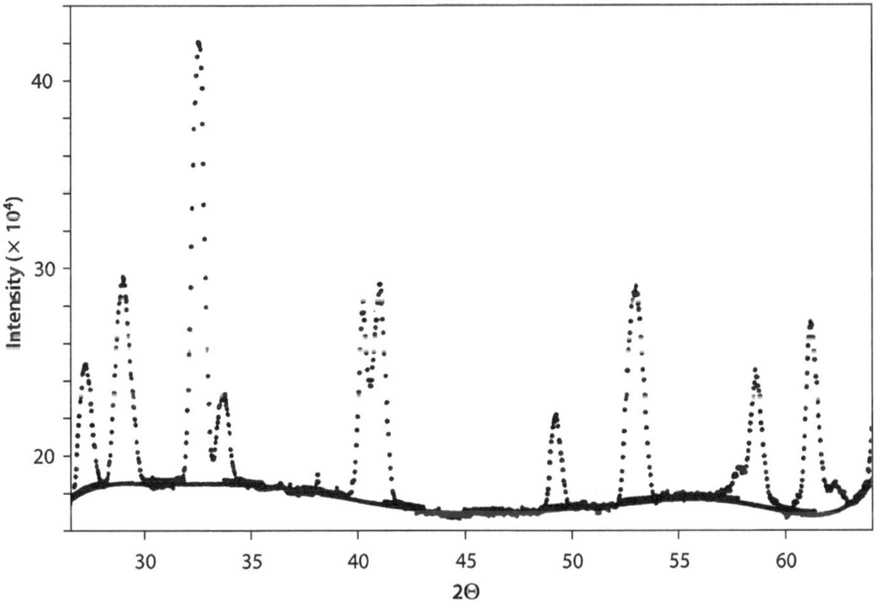

Fig. 3.44. Determination of background by fitting a polynomial, here of degree nine. The specimen was ZnAl alloy

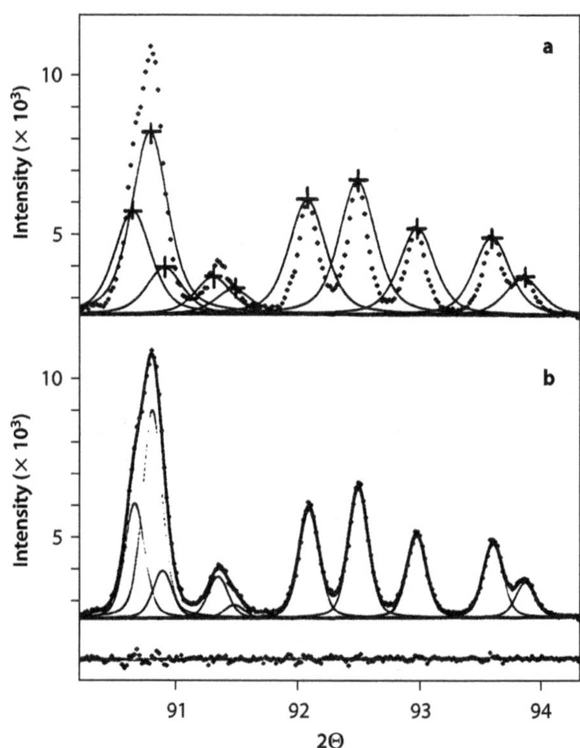

Fig. 3.45. a Expected peak positions marked by cross hairs; **b** result of the profile fitting

the central peak decreases at the expense of the two maxim at a pole distance of about 30°, indicating that the 100 planes are forced perpendicular to the rolling direction.

3.9.5.2
Rocks

Numerous rock samples were studied, in particular quartzite, hematite ores, pyrite or chalcopyrite In general in those specimens 864 data sets were recorded consisting of between 11 and 30 reflections. The total time for one set of measurements was typical between 35 minutes for pyrite and 95 minutes for chalcopyrite, where 33 reflections were recorded and processed. Figure 3.48 gives two examples of chalcopyrite samples showing the sum patterns after profile analysis. The upper pattern is a two-phase sample with chalcopyrite (K) and pyrite (P), the lower pattern has three phases: chalcopyrite (K), pyrite (P) and magnetite (M). From the diffraction data pole figures are calculated, where Fig. 3.49. gives such pole figures (101)/(011), (110) and (102)/(012) of six different quartzite samples. Figure 3.50 gives the corresponding ODF representation of one sample, BGR 420, calculated from the data of Fig. 3.49.

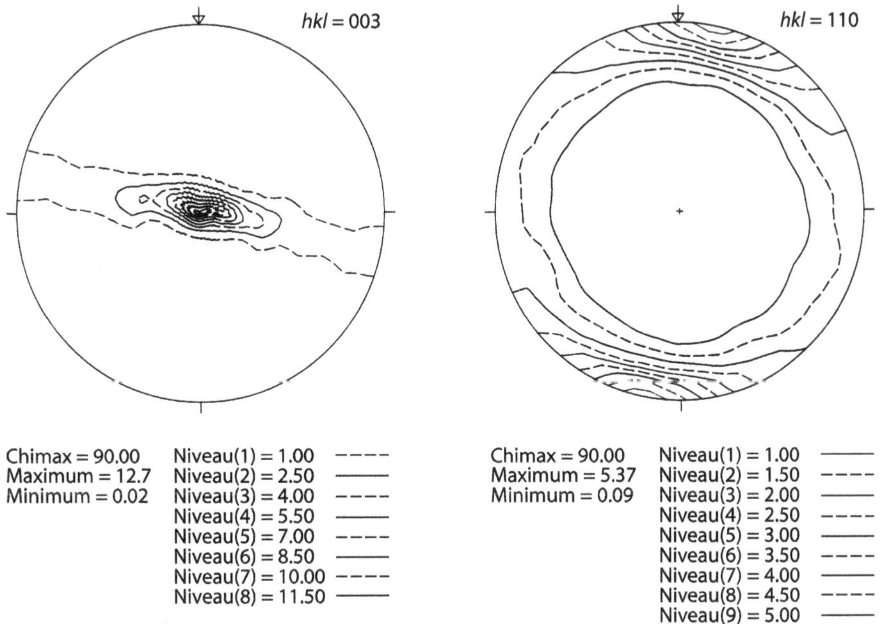

Fig. 3.46. Two pole figures in stereographic projection of hematite, extracted from data shown in Fig. 3.42. Shown are the projections (003) and (110). The hematite crystals in this sample are highly oriented along the c-axis

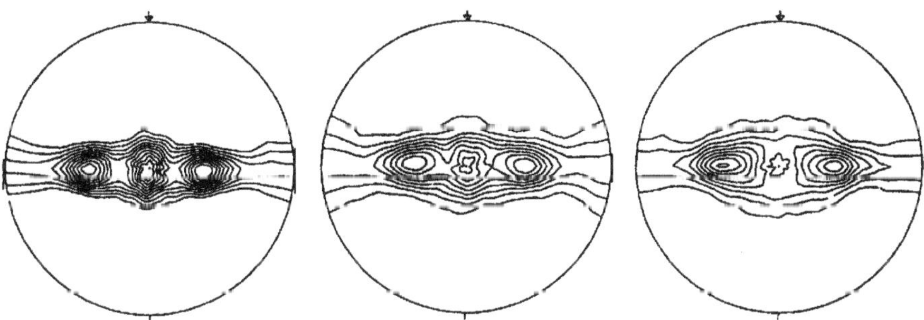

Fig. 3.47. 002 pole figures of cold rolled titanium metal sheets at different rolling degrees: 40% (*left*), 50% (*middle*), 80% (*right*)

3.9.5.3
Meteorites

Interesting examples of texture analysis are studies of meteorites. There are very few known investigations. An important feature in using neutron diffraction in studying

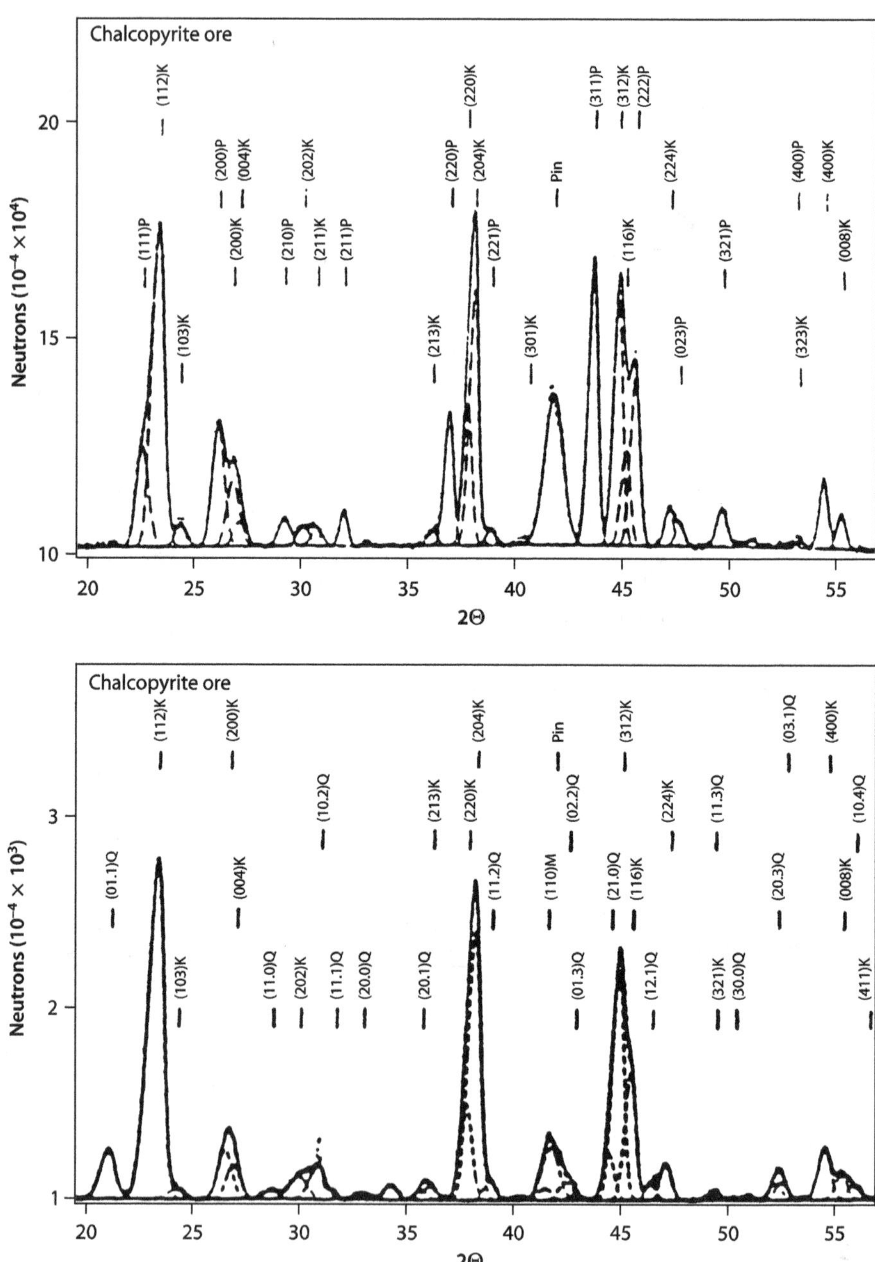

Fig. 3.48. Diffraction patterns of two chalcopyrite samples after adding up all individual single position measurements ("sum patterns"). The *upper picture* is a two-phase sample with K = chalcopyrite and P = pyrite, the *lower picture* is a three phase sample with K = chalcopyrite, Q = quartzite and M = magnetite

Fig. 3.49. Pole figures of six quartzite samples: (101)/(011), (110) and (102)/(012)

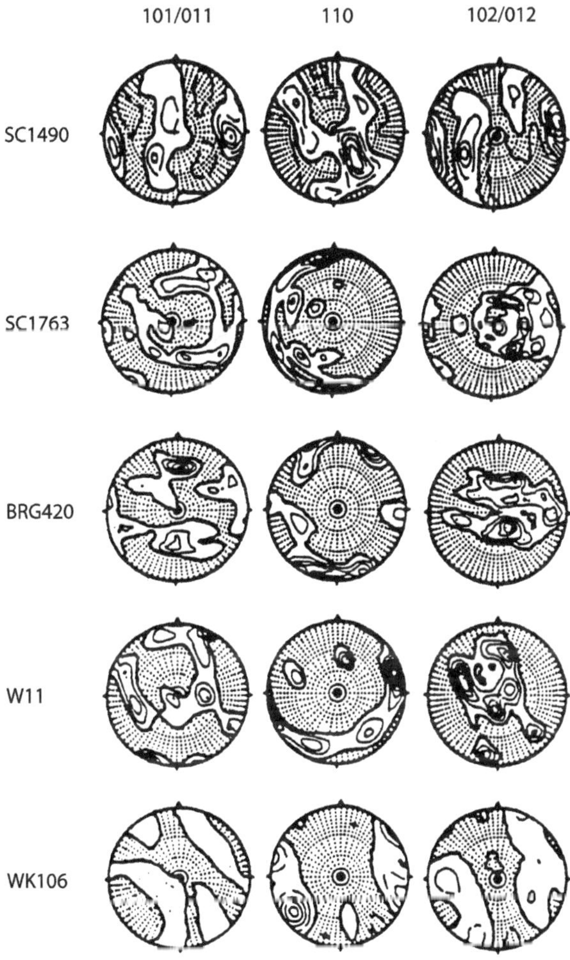

meteorites is the fact, that the experiments do not require cutting of the general precious samples. Reports of this kind come from the Bonn University Group in Jülich (Höfler et al. 1988, 1989; Will et al. 1988b). They studied three samples from the museum collection: the hexahedrites "Coahuila" and "Walker County", both belonging to Group IIA, and the octahedrite "Gibeon", Group IVA (Höfler et al. 1988). An interesting and important feature is that these samples are believed to be fragments of single crystal meteorites, so we do not really investigate texture in its true meaning, which is the preferred orientation of crystallites in a polycrystalline sample, but rather the orientation of crystallites, broken up from a single crystal after phase transformation in the process of cooling, or after mechanical twinning caused by shock events on entering the atmosphere of the earth.

Fig. 3.50. ODF representations for BGR 420 calculated from the data from Fig. 3.49

BRG420

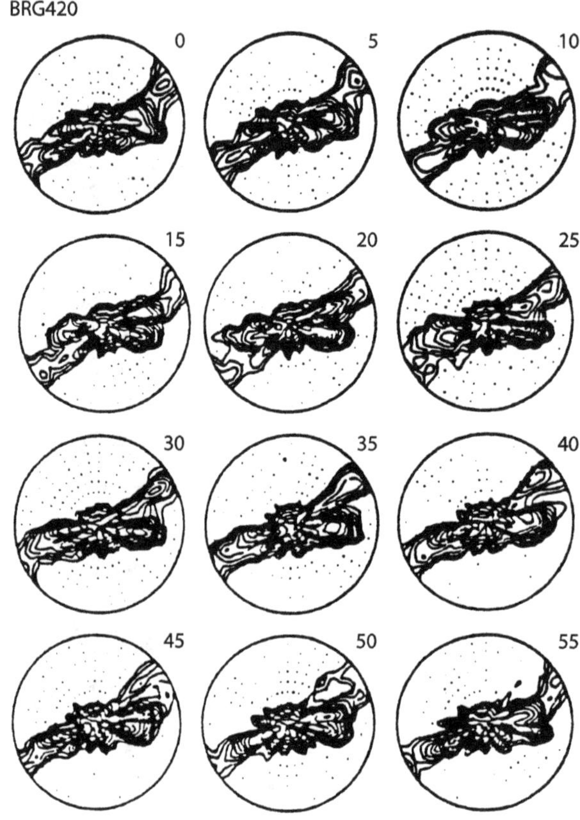

Hexahedrites are iron-nickel meteorites composed of large pieces of single crystals of kamacite, a bcc α-Fe,Ni alloy with less than 6% Ni. Because of the single crystal nature of such meteorites sharp peaks can be expected. This requires smaller increments than usual, here $\Delta\phi = 1°$ at the outer part of the pole figure and up to 4° for $60° < \chi < 90°$. The increments in pole distance were $\Delta\chi = 2.5°$. This fine grid of measurements led to much larger data sets with a total of 7 561 diffraction patterns. With 12 s per step (= increment) the total time spent for one specimen was 25 hours. Following the standard procedure the data were normalized and pole figures were calculated and plotted in stereographic projection. On the hexahedrite specimens pole figures of the kamacite reflections 200, 110 and 211 were measured. Figure 3.51 gives a typical example from the Walker County meteorite. All pole figures from the three meteorites investigated are characterized by sharp density pole peaks surrounded by very low background. The result is basically a projection of single crystals. The peaks with large extension and high intensity are the face poles of a cubic single crystal. This can easily be recognized in the pole figure 200 where three such poles are to be seen. Peaks with low intensity and small extension are believed due to 211 twinning. Figure 3.52 shows the same pole figure with the result of calculated pole sites according to the (221) twin mirror planes. The single crystal nature of the meteorites is demonstrated in Fig. 3.53, which shows the ϕ-scan of such a pole density peak from the Walker County pole figure 200 at $\chi = 12.5°$. These investigations show the power of texture analysis in combination with profile fitting especially for the study of meteorites.

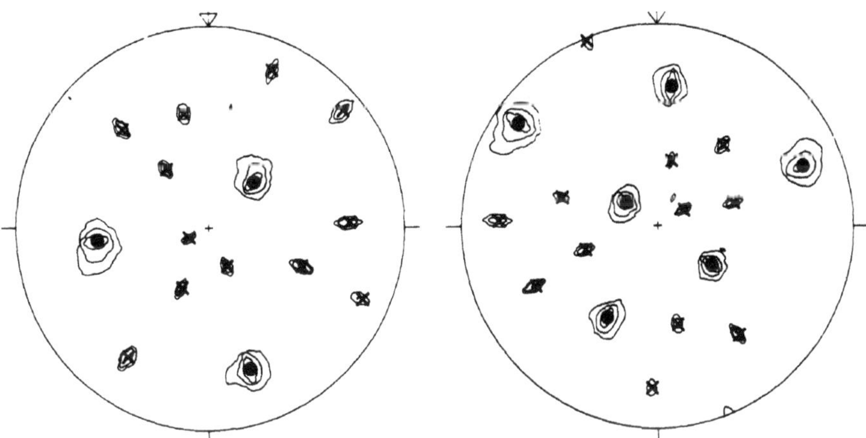

Fig. 3.51. Pole figures 200 and 110 of the meteorite from Walker County as determined from neutron texture analysis

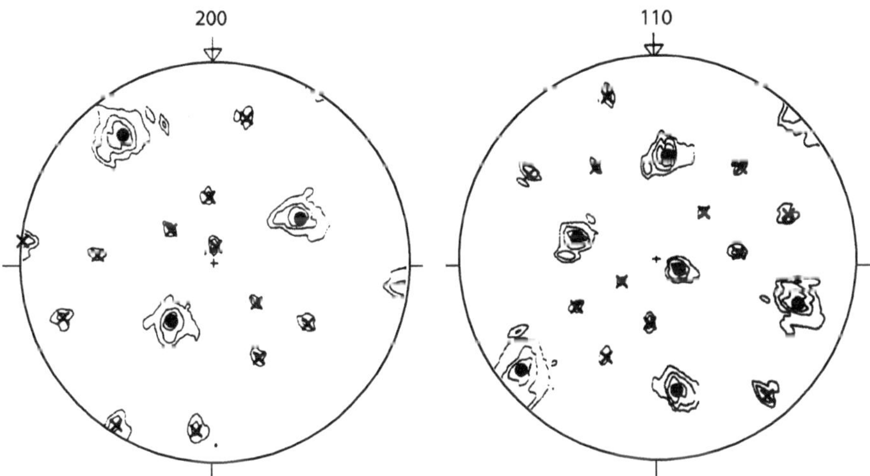

Fig. 3.52. The same pole figures 200 and 110 of Fig. 3.51 now including the calculated pole sites according to the (221) twinning. Single crystal face poles marked with *dots*, twin face poles marked with *x*

Summing up this investigation of meteorites Fig. 3.54 shows the 200 pole figure of camacite from the meteorite Gibeon (left), and the same pole figure including the analysis of the calculated α-phase (100) poles calculated according to Kurdjumow and Sachs (1930). The agreement is convincing. An analysis based on the orientation relationship suggested by Nishiyama (1934) and Wassermann (1935) gave less agreement. For details see Höfler et al. (1988).

3.10
Selected Examples for the Application of the Two Stage Method

The Two Stage method has been used in numerous applications. Structure refinements with data from X-ray, synchrotron and neutron diffraction patterns were the main applications. The intensive refinement of quartz data has been discussed in detail in Sect. 3.8. Some additional specific examples shall be added to demonstrate the powerful use of this method:

- Electron density calculations by Fourier methods of an olivine sample
- Electron density distribution in CeO_2
- Crystal structure analysis of Yb_2O_3, a study of anomalous dispersion
- Application in high pressure research
- Cation distribution in a thin film garnet sample

3.10.1
Electron Density Distribution of an Olivine Sample by Fourier Methods

The full power and advantage of the Two Stage method is realized for *ab initio* crystal structure determination, for example if Patterson diagrams are needed, or if we want to calculate electron density maps, or even with so-called direct methods. This can be done without difficulties in Step 2 of the method. As an example we show the results from Mg_2GeO_4, a material homologue to the mineral olivine, important in earth sciences (Will et al. 1988b). Natural olivine, $(Mg,Ge)_2SiO_4$, is a major component in the earth mantle. It undergoes a phase transformation at the upper mantle to mantle boundary under pressure and temperature from the orthorhombic olivine structure to the denser cubic spinel structure and therefore has attracted much interest and many investigations. However the high pressures and high temperatures required for these transformations are outside routinely available diffraction conditions and therefore defy experimental studies of this transformation. The germanate-olivines exhibit the same phase transition behavior as the Si-olivine, with however much lower pressures and temperatures (Will and Lauterjung 1987). While for natural olivine single crystals are readily available, no single crystal could be grown for the germanate homologue.

Mg_2GeO_4 crystallizes orthorhombic in space group Pbnm with $a = 4.9106(3)$, $b = 10.3214(6)$ and $c = 6.0365(3)$ Å. The diffraction pattern was collected with synchrotron radiation with the rather long wavelength of $\lambda = 1.74$ Å (for better peak separation) and a 2Θ range of 18–85°. The step width was 0.01°, and $t = 2$ s per step giving a total run time of about 4 hours. With a greatly increased resolution due to an improved long parallel slit system and a longer wavelength most peaks were well resolved and

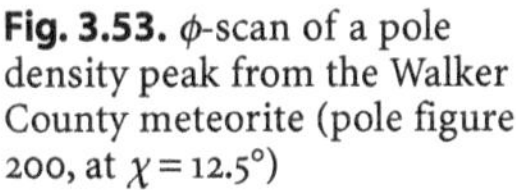

Fig. 3.53. ϕ-scan of a pole density peak from the Walker County meteorite (pole figure 200, at $\chi = 12.5°$)

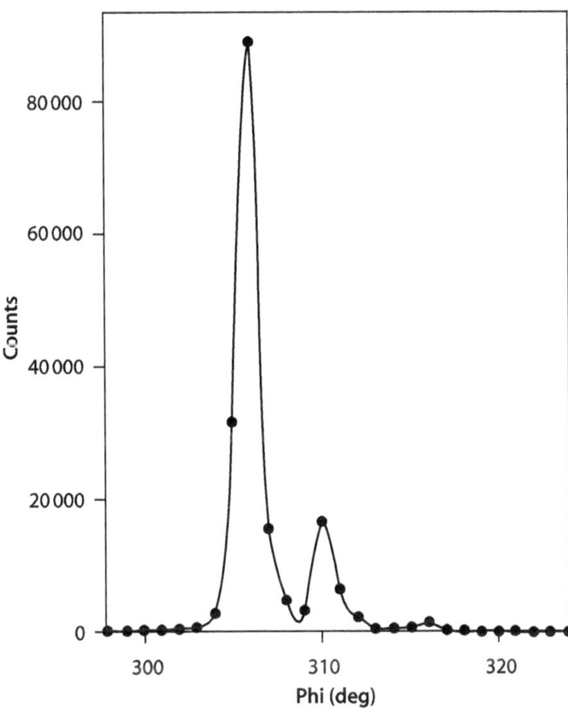

Kamacite 200

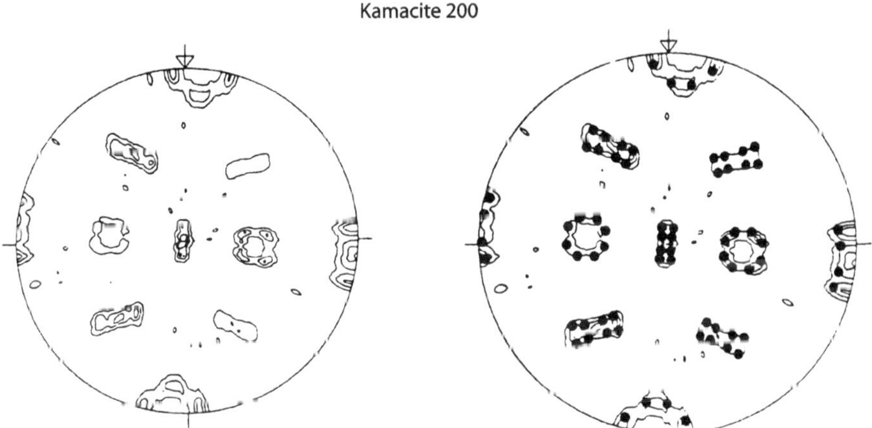

Fig. 3.54. Pole figure 200 kamacite of the Gibeon meteorite (*left*), and the calculated orientation relationship according to Kurdjumow and Sachs (1930) (*right*)

the others could easily be separated with profile fitting. In all 81 reflections were obtained by profile fitting. Figure 3.55 depicts the diffraction pattern, Fig. 3.56 shows two typical sections of the pattern.

In cases like this one with a rather large unit cell and the lattice parameters known or determined beforehand, possibly from well resolved reflections in the forward region of the pattern, it is advisable to perform a full pattern profile fitting keeping the lattice parameters constant and fixed and therefore also the peak positions fixed. This was done in this case of Mg_2GeO_4 with the program FULFIT.

With this investigation we were able to go well beyond a purely structure refinement to an actual structure analysis. The profile fitting calculations gave a list of integrated intensities together with the *hkl* values. Because of the orthorhombic symmetry there is no intrinsic overlap of reflections and as a result we have a set of observations, integrated intensities first and after correcting for Lorentz effects a set of structure factors $|F|$.

This is the basis for calculating Fourier maps, e.g. electron density distributions directly from observed powder data. Since the structure type is known, the olivine structure, the phases, e.g. because of centrosymmetry only the signs +/−, need to be calculated and applied to the observed structure factors. Fourier sections are shown in Fig. 3.57a with the plane containing O1-Ge-O2 of the GeO_4 tetrahedron in the left picture. The Mg atom is above the plane, but can still be seen at the right top and bottom corners. In Fig. 3.58 a section through the GeO_6 octahedron is shown. Visible are the Mg ion, four oxygen atoms and two Ge ions.

It was possible in this case to make a direct comparison between powder diffraction data and single crystal data of the isostructural natural Si-olivine, shown in

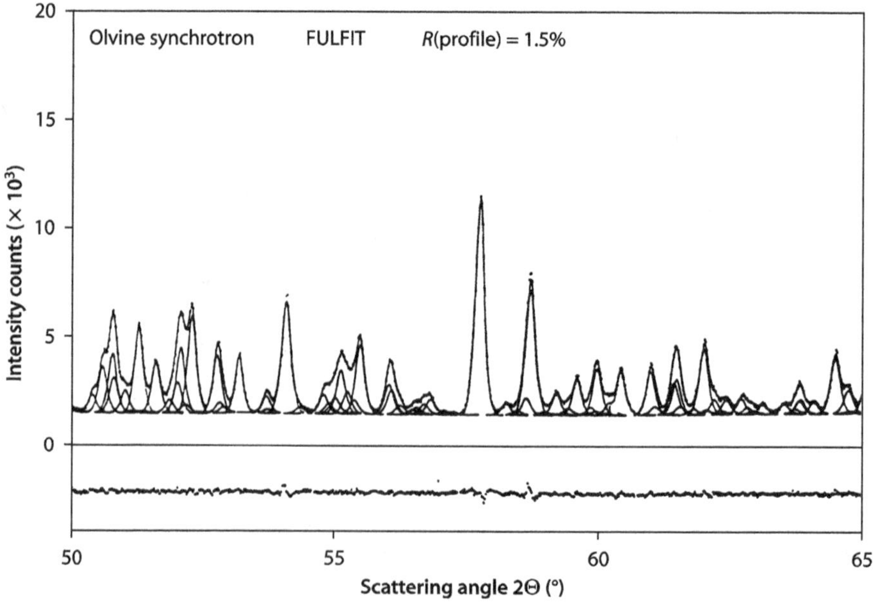

Fig. 3.55. Diffraction pattern of the orthorhombic Mg_2GeO_4 olivine. Section from $2\Theta = 50$ to $65°$

Fig. 3.57b. The single crystal study contained 1 349 reflections, much more than was possible with powder diffraction. Therefore for a fair comparison we have taken from the single crystal study the same 81 reflections. As to be expected the electron density at the silicon site is smaller than for germanium because of the difference in absolute

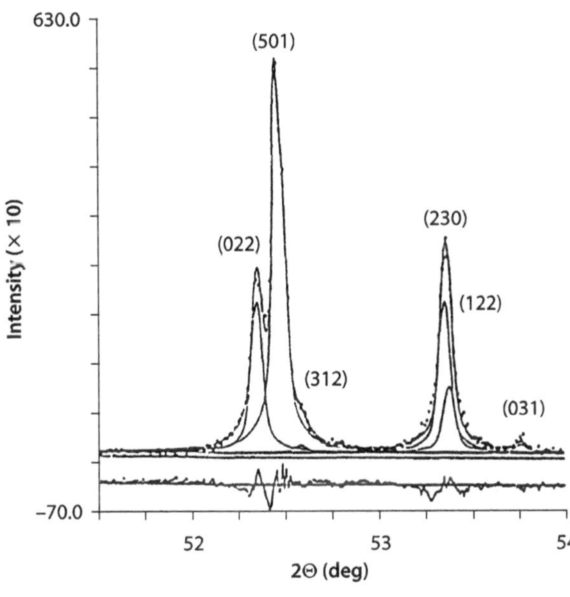

Fig. 3.56. Two profile fitted sections of Mg_2GeO_4. Differences between experimental and calculated points are shown below each section. The resolution is 0.05°

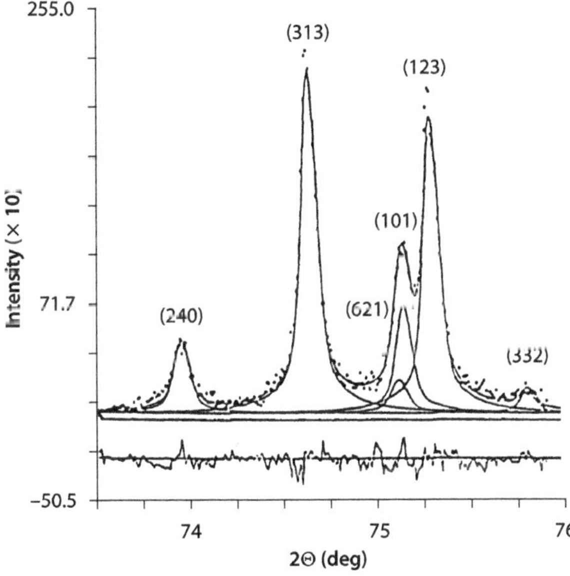

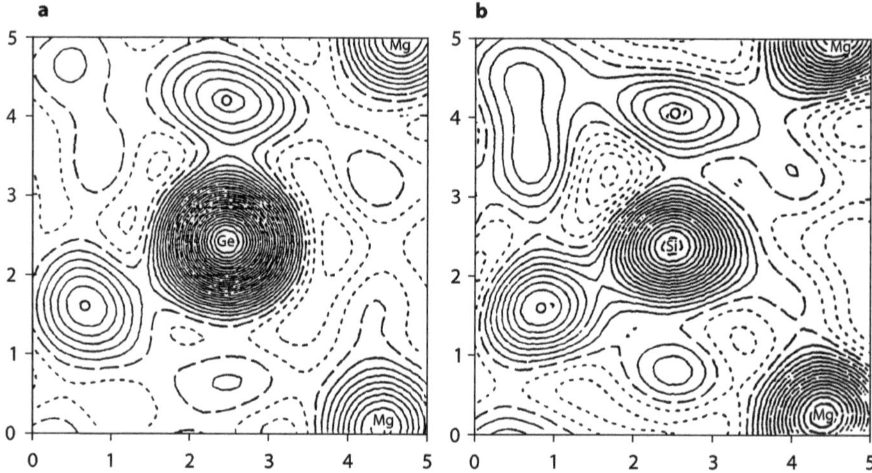

Fig. 3.57. a Fourier map of a section of Mg_2GeO_4 through the GeO_4 tetrahedron calculated with the structure factors obtained directly from the profile-fitted powder diffraction data; **b** gives the analogue section for the natural olivine $(Mg,Fe)_2SiO_4$ calculated with single crystal data. Here we have silicon instead of germanium

Fig. 3.58. Similar Fourier map as in Fig. 3.57, now through the GeO_6 octahedron

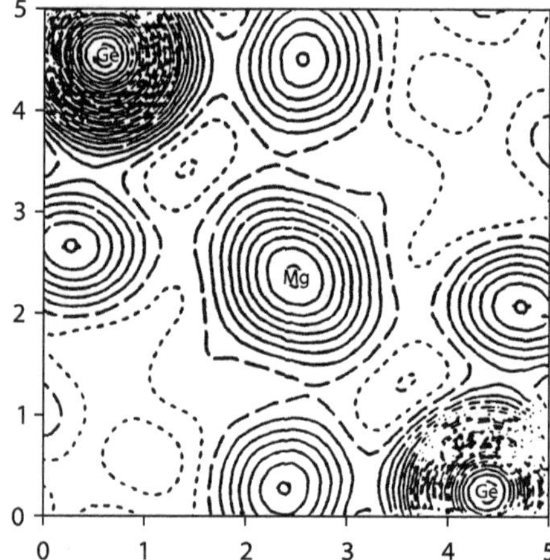

electrons in both elements. The excellent agreement between the two maps is a convincing demonstration of the power of the Two Stage method and the possibilities available with powder diffraction.

3.10.2
Electron Density Distribution in CeO$_2$

This example gives another very a convincing application of the Two Stage method. Here the investigation was based on a separation of data in HO (high order) and LO (low order) regions. CeO_2 is a simple compound crystallizing in the CaF_2 type structure with no positional parameters. The refinement of the data should therefore yield very low R-values. CeO_2 was measured with synchrotron radiation giving a very good diffraction pattern, as is shown in Fig. 3.59 well resolved peaks, high intensity and low background. The conditions were $\lambda = 1.0$ Å, $\Delta(2\Theta) = 0.02°$, $t = 4$ s. Preferred orientation was absent in this specimen. The least squares refinement with POWLS with 32 partly intrinsically overlapping peaks of the 52 possible hkl values did not refine below $R = 2.64\%$. This is better than a Rietveld refinement of CeO_2 published by Cox et al. (1983) with $R = 6.3\%$, but with the high quality data available and from our experience it is not acceptable. Looking at the differences between observed and calculated intensities in the POWLS calculation gave very good agreement for high angle reflections, but significant differences in the low angle region. This is depicted in Fig. 3.60 and in Table 3.19. The obvious conclusion is that the compound, believed to be a simple ionic crystal, must have appreciable covalent, delocalized electron densities between the ions. As a consequence we treated the data in the general method of chemical bond-

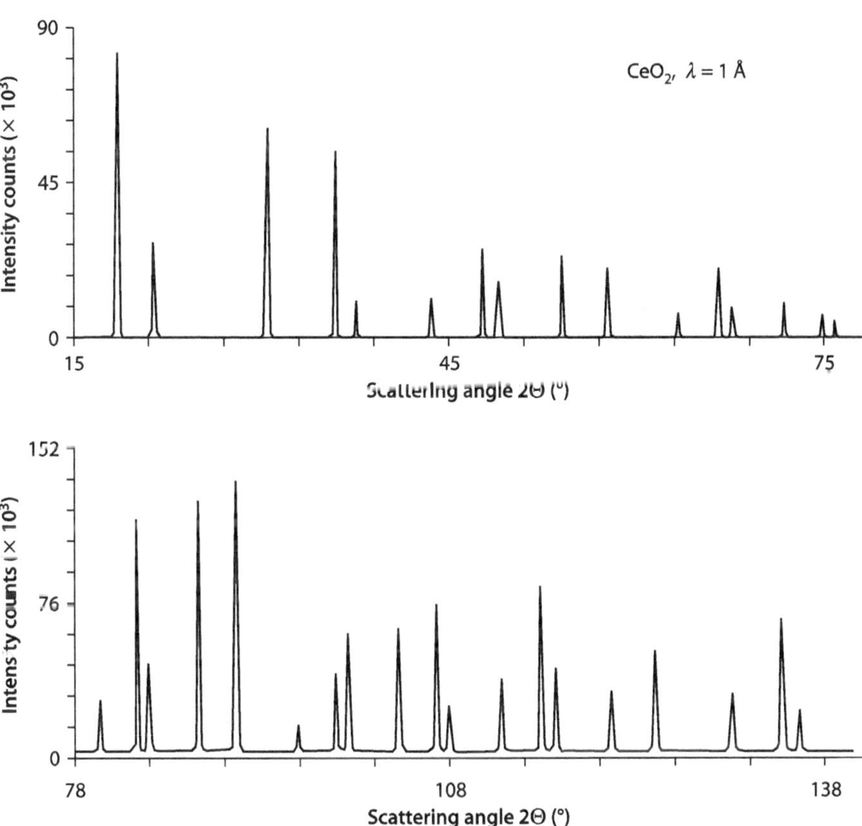

Fig. 3.59. Powder diffraction pattern of CeO_2 recorded with synchrotron radiation. The intensity scale of the lower pattern was decreased to show weaker reflections more clearly

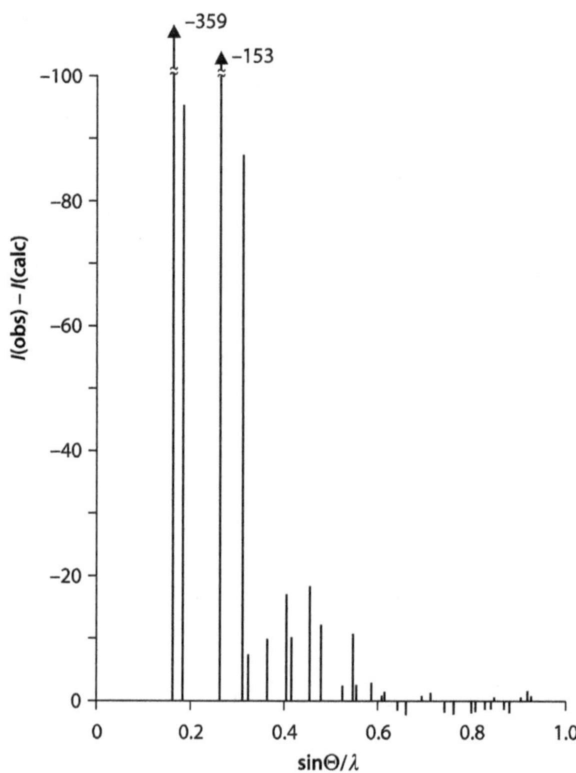

Fig. 3.60. Differences between observed and calculated intensities of CeO_2 after refinement with POWLS. There are large differences in the low angle region

Table 3.19. Comparison between observed and calculated intensities of CeO_2 after refinement with POWLS. Listed are only the first ten LO reflections, which have large discrepancies between observed and calculated values

hkl	*I*(obs)	*I*(calc)	*hkl*	*I*(obs)	*I*(calc)
111	1 641.60	1 968.77	400	210.17	218.58
200	520.23	602.18	331	480.58	494.11
220	1 118.43	1 262.56	420	318.44	325.21
311	1 021.17	1 099.04	422	446.01	461.49
222	203.36	209.12	511,333	376.40	386.59

ing investigations by separating the data set into high order (HO) and low order (LO) sections (see for example Coppens 1971; Kirfel and Will 1980).

Rewriting and adapting the program POWLS to separate HO and LO data in the refinement yielded a R-value of $R = 0.55\%$ for 21 observations with indices higher than $s = \sin\Theta / \lambda = 0.6\ \text{Å}^{-1}$ (Table 3.20). The next step is to calculate difference Fourier maps $F(\text{obs}) - F(\text{calc})$. One result is shown in Fig. 3.61 depicting the (110) section, with ce-

Table 3.20. Comparison between observed and calculated intensities of CeO_2 after refinement with POWLS. Included in the refinement are only the HO reflections. The R-value is 0.55%

hkl	I(obs)	I(calc)	hkl	I(obs)	I(calc)
533	158.41	158.29	751,555	164.18	163.69
622	116.99	117.75	662	55.10	53.63
444	55,66	55.37	840	83.28	82.05
551,711	245.00	244.10	753,911	188.35	188.15
640	93.29	91.94	842	97.70	97.35
642	266.64	266.41	664	75.09	74.12
731,553	291.23	293.44	931	118.50	118.39
800	27.10	27.04	844	71.69	72.54
733	80.79	81.12	771,933	175.20	176.20
820,644	124.13	123.44	755		
822,660	139.22	138.65	860,10.0.0	57.56	58.26

rium at 000 and oxygen at ¼,¼,¼. There is considerable charge accumulation between two cerium atoms with corresponding electron deficiencies at the cerium sites. The oxygen positions are not affected. A first picture would suggest that electron charges are moved from the cerium into the bonding between two Ce atoms, a picture which is at the moment difficult to understand. As is customary in such investigations we put point charges, with corresponding spatial expansion into those sites. It improved the R-value of the total data set from $R = 2.6\%$ to $R = 1.7\%$, a significant improvement, but still not describing the full picture and the bonding behavior in this compound. This investigation obviously leaves some questions open.

3.10.3
Crystal Structure Analysis of Yb_2O_3

Yb_2O_3 was studied in order to determine the anomalous dispersion coefficients f' and f'', which are needed in protein crystallography for phasing. Near the absorption edges, when the energy of X-rays becomes comparable to the absorption levels of the atoms in the crystal, the scattering is affected in amplitude as well as in phase. Anomalous X-ray scattering is well-known as a powerful tool for estimating the phases in crystal structure analysis. This method is used primarily in protein structure determinations, where rare-earth atoms showing large anomalous scattering effects at their L-edges are especially well suited. Using this method for phase determination requires selecting a wavelength close to an absorption edge. Measurements at several wavelengths are advisable, and this is possible with synchrotron X-rays. Relativistic calculations of anomalous scattering factors were published by Cromer and Liberman (1970, 1981). However, these calculations generally do not show sufficient agreement with experimental data for the L absorption edges. Despite corrections and improvements of the theory (Liberman, personal communication in 1986), there are still systematic differ-

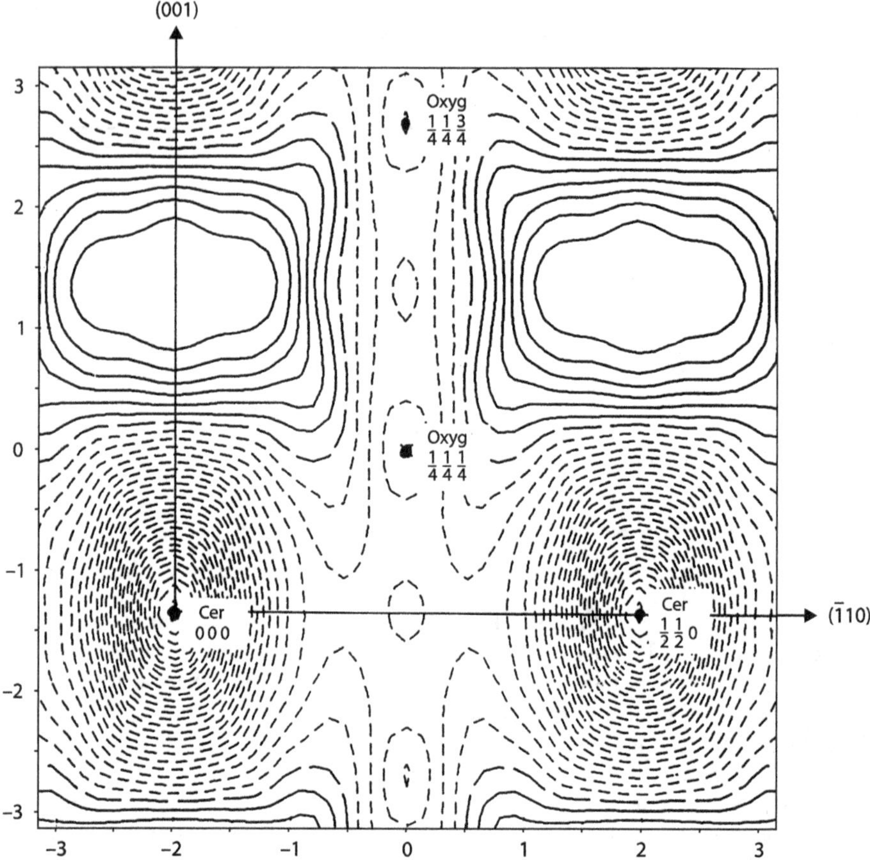

Fig. 3.61. Difference Fourier map (110) of CeO_2 calculated with the structure factor differences from the refinement with POWLS. Contours are 0.1 el Å^{-3}

ences between measured values and those calculated. The inadequacy of the theoretical model is evident and these terms therefore have to be determined experimentally. If it is done by the usual single crystal work it is time consuming requiring long measuring times at a synchrotron. Therefore powder diffraction would be most suitable, and this method was tested in an investigation described here (Will et al. 1987a). Crystal structure information can be obtained with considerable confidence and precision from synchrotron radiation powder data meeting the requirements for this goal. With the Two Stage method and using the refinement program POWLS overlaps in powder diffraction patterns can be handled.

Using synchrotron radiation we have the possibility to choose different wavelengths on a routine basis, and especially close to the absorption edge. Yb_2O_3 was selected as a representative example in cooperation with Prof. Liberman, who provided theoretical values. Oxygen served as an "internal standard" so that the absolute scale factor is totally independent of the f' values. Data were collected with four wavelengths slightly longer than the Yb-L_{III} edge defined in Fig. 3.62. The absorption changes rapidly in the vicinity of the absorption edge and is very large on the short wavelength side. An important advantage of the symmetrical powder diffractometer geometry used in these experiments is that the incident and reflected rays make the same angle to the powder specimen surface at all values of 2Θ. The integrated intensities are proportional to $1/\mu$ where μ is the linear absorption coefficient of the specimen and are independent of 2Θ so that no corrections are required.

With this technique only the values of f' for Yb can be determined experimentally, the correction term in the scattering factor of Yb at the L_{III} absorption edge. Owing to the intrinsic overlaps of Friedel pairs (hkl) and $(-h,-k,-l)$ in powder diagrams, the imaginary term f'' cannot be derived from the intensities. However f'' can be calculated from f' by the Kramers-Kronig dispersion relation.

Data were collected at the Stanford Synchrotron Radiation Facility (SSRL) with a Si(111) channel monochromator and parallel slit geometry. The diffraction pattern collected with $\lambda = 1.3895$ Å is shown in Fig. 3.63. The angular range was arbitrarily limited to $2\Theta = 14$ to $88°$ yielding 48 well resolved peaks from 101 possible (hkl) planes partly intrinsically overlapping at exactly the same 2Θ position due to the crystal structure symmetry having the same $N = h^2 + k^2 + l^2$, but with different integers or sequence of integers. Step width was $\Delta(2\Theta) = 0.02°$, counting time was 1 s per step. Preferred orientation was checked with corrections between 0 and 3.6%. Yb_2O_3 crystallizes in the α-Mn_2O_3 type structure with the atomic positions shown in Table 3.21.

The lattice parameter was 10.436(1) Å as determined from least squares refinement. The integrated intensities were determined by profile fitting methods. The best fitting function in this specific case was achieved by superimposing two Gaussian functions with a 1:2 ratio of the *FWHM* and the peak height. A typical profile fitted section is shown in Fig. 3.64, where the differences of observed and calculated points are shown at half height with the same intensity. In the structure studied here there were no overlapping peaks, but a large number of intrinsically superimposed peaks with different

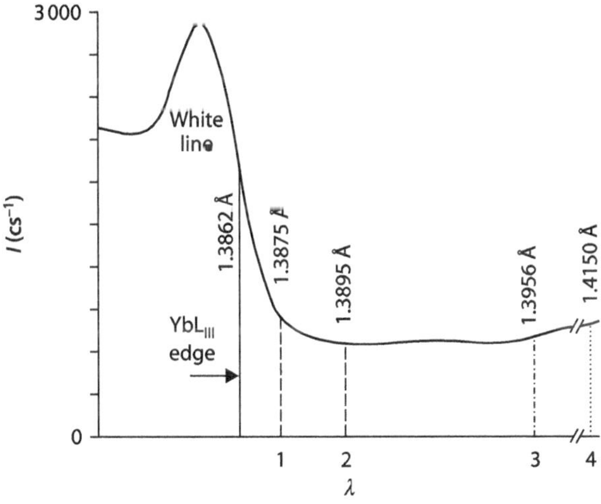

Fig. 3.62. Spectrum in the vicinity of Yb-L_{III} absorption edge obtained with energy-dispersive diffraction and Yb_2O_3 powder sample. The four wavelengths used are indicated by the *dashed lines*

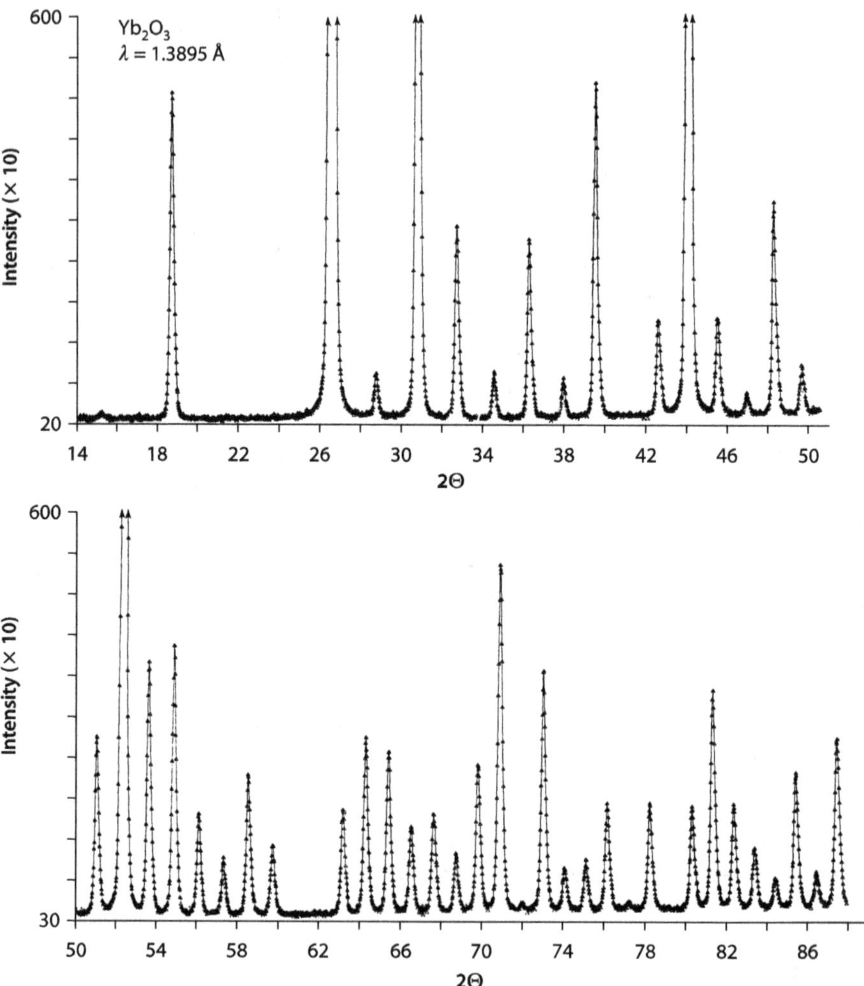

Fig. 3.63. Synchrotron powder diffraction pattern of Yb_2O_3 for $\lambda = 1.3895$ Å. The highest peak intensity was 56 600 counts s^{-1} for (222) at 26.7°

Table 3.21. Atomic positions for Yb_2O_3 crystallizing in the α-Mn_2O_3 type structure

Yb(1)	8(b)	¼	¼	¼
Yb(2)	24(b)	x	0	¼
O	48(l)	x	y	z

intensities, as required by the space group. The refinement contained 48 diffraction peaks with 101 (hkl) Miller planes. The refinement included the positional parameters x, y, z for oxygen and x for Yb(2), as well as individual isotropic temperature factors. Table 3.22 gives a compilation of the refined structural data.

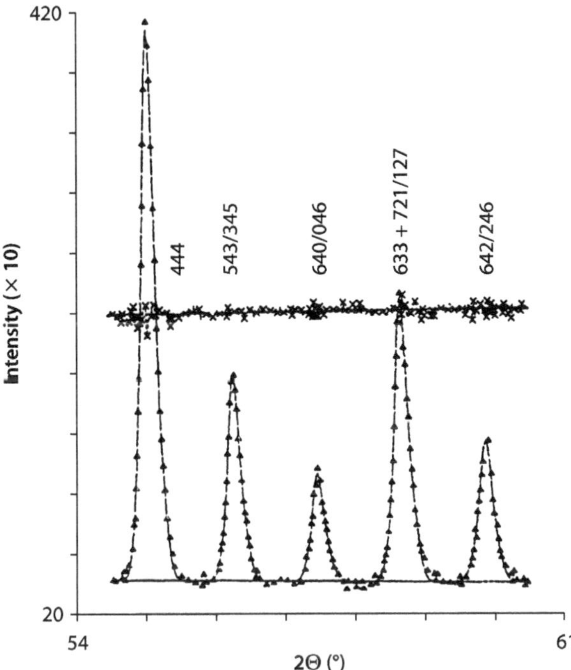

Fig. 3.64. Profile fitting with double Gaussian functions of five weak reflections. $R_{PF} = 1.2\%$

The method is based on a good knowledge of the crystal structure parameters, which can best be derived with a wavelength where f' is negligible, or at least very small. The investigation therefore started with a wavelength far from the absorption edge. For this data set f' was varied systematically in a series of least squares calculations until the lowest R-value was obtained. Since the measurements were made on the soft side of the absorption edge, there is no contribution from the L-shell to the imaginary correlation term f''. f' was assumed to be independent of scattering angle. The final results are shown in Table 3.22.

Figure 3.65 shows the effect of wavelength on the position, intensity and background for two reflections, 400 and 844. As the wavelength decreases approaching the Yb-L_{III} absorption edge, the relative intensities decrease because f' is negative and large. The fluorescence background increases. The intensity increases with increasing contributions of f', as we move away from the edge. This is also demonstrated in Fig. 3.66, where the form factor of Yb^{3+} is plotted together with the contribution from f'. The rate of increase is greater for higher diffraction angles, because the intensity is determined basically by the value of $f + f'$ which is $f - |f'|$ because f' has the opposite sign to f and also as a consequence of a strong fall-off of the form factor f with $(\sin\Theta)/\lambda$. The effective scattering powers $f(\lambda) = f_0 - f'(\lambda)$ are shown for $\lambda = 1.4150$ Å for the two reflections 400 and 844. f' is taken to be independent of $(\sin\Theta)/\lambda$.

The experimentally determined values of the real part of the anomalous correction term f' for Yb are 3.5 to 5.1 electron units higher than the theoretical values calculated from the Cromer and Liberman (1970, 1981) theoretical method for K- and L-edges. These differences are believed to be real and are similar to those obtained for

Table 3.22. Refined structure data for Yb_2O_3

	Wavelength (Å)			
	1.3875	1.3895	1.3956	1.4150
Data set	1	2	3	4
f'(exp)	−21.2(2)	−19.5(2)	−15.5(2)	−13.7(2)
f'(theor)	−16.41	−14.42	−12.19	−9.86
Difference	4.8	5.1	3.3	3.8
a_0		10.436(1)[a] *10.4322(5)*[b]	−0.0321(1)	−0.0322(1)
x(Yb)	−0.0322(1)	−0.0322(1) *−0.03253(4)*[b]	0.3909(12)	0.3912(11)
x(O)	0.3905(11)	0.3918(9) *−0.3910(6)*[b]	0.1556(10)	0.1545(9)
y(O)	0.1551(9)	0.1545(8) *−0.1523(6)*[b]	0.3796(13)	0.3802(11)
z(O)	0.3798(12)	0.3805(9) *−0.3807(6)*[b]	0.24(8)	0.15(6)
B(Yb1)	0.20(9)	0.12(6) *−0.25*[b]	0.11(4)	0.08(3)
B(Yb2)	0.06(5)	−0.03(3) *−0.21*[b]	0.28(18)	0.24(17)
B(O)	0.30(17)	0.48(18) *0.49*[b]	0.067(10)	0.058(8)
GP[c]	0.061(10)	0.058(8)	1.65	1.42
R_{Bragg} (%)[d]	1.75	1.26 *3.52*[b]		

[a] Average of four wavelengths.
[b] Values in italics from single crystal data (Saiko et al. 1985).
[c] Preferred orientation plane (111).
[d] Using unit weights.

Sm and Gd by Templeton et al. (1980, 1982) using single crystal methods. The experimental and theoretical values f' calculated by the Templetons are plotted in Fig. 3.67.

3.10.4
Application in High Pressure Research

High pressure research when using synchrotron radiation is an especially sensitive application for the Two Stage method. Because of the specific circumstances the Rietveld method cannot be, and has not been applied. Those problems are:

- There are always peaks from several compounds, in all cases the sample itself, a pressure marker, NaCl, MgO or gold for example, and the metal gasket containing the specimen

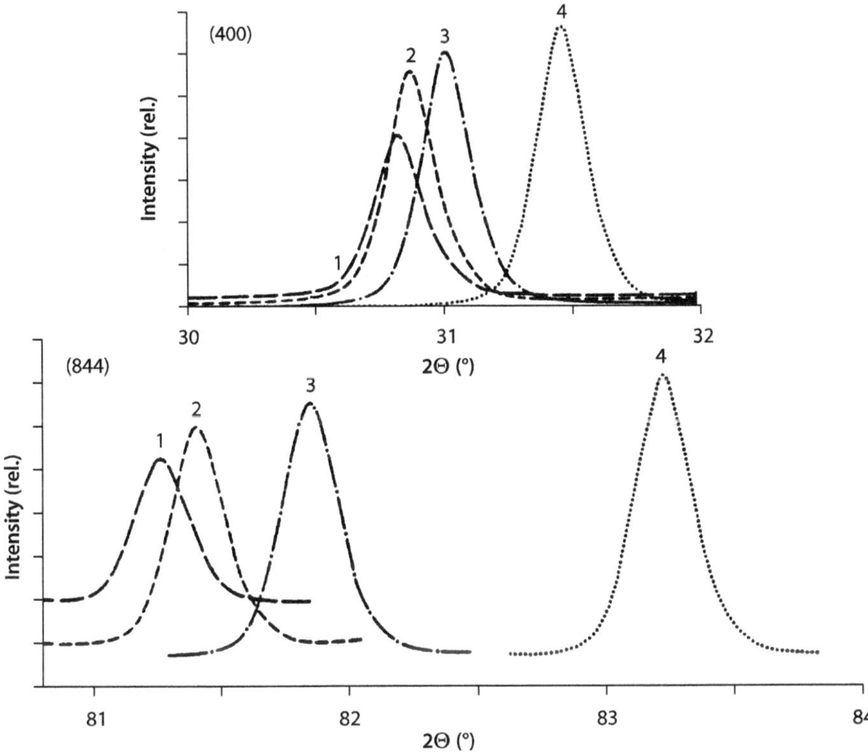

Fig. 3.65. Effect of wavelength on intensity and background of Yb_2O_3 for the 400 and 844 reflections

- The pattern contains many peaks very close to each other, which make it difficult to place footing marks for possible background subtraction
- The background is hard to describe in a general way. The program developed by Lauterjung (Lauterjung et al. 1985) using polynomials of order up to five, as suggested by Steenstrup (1981) is a reasonable technique to determine the background. It has been applied successfully on many occasions
- In energy dispersive mode, the technique normally used, the background, and also the intensity of the specimen may go through an absorption edge
- In most diagram escape peak are present, which have to be treated, i.e. excluded manually

As a representative example a high pressure study of CuS_2 is mentioned here briefly (Hüpen et al. 1991). Data were collected at the HASYLAB at DESY in Hamburg. MgO was used as a pressure marker. Figure 3.68 shows the diffraction pattern collected at $p = 24$ kbar. The exposure time was 500 s. The maxima are labeled accordingly to CuS_2, MgO and the gasket (G). The pattern is characterized by many strongly overlapping peaks. They could be separated by profile fitting. Figure 3.69 shows for this purpose

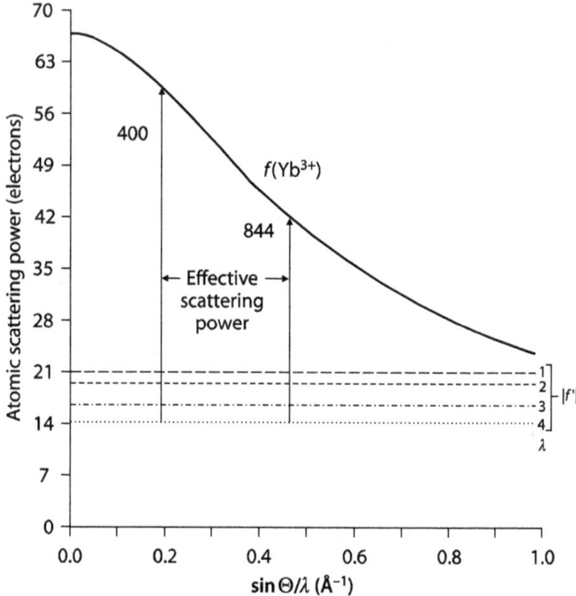

Fig. 3.66. The scattering factor $f(\text{Yb}^{3+})$ and the contribution of the anomalous scattering term f'

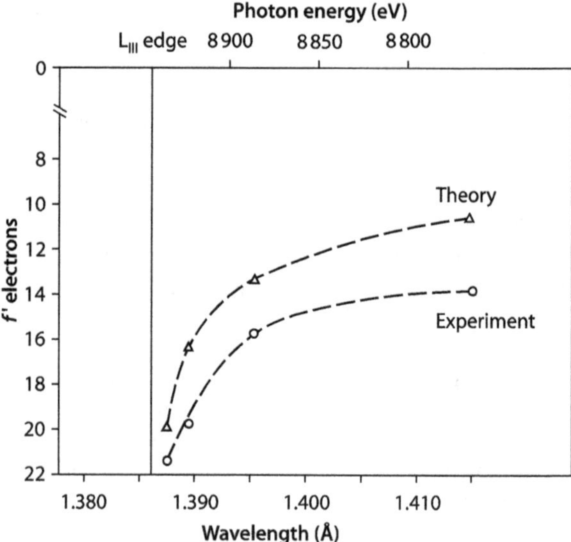

Fig. 3.67. Experimental and theoretical values of f' for Yb_2O_3 near the Yb-L$_{\text{III}}$ absorption edge as a function of energy

the measured raw pattern without any corrections, with the insert giving the high energy/low intensity region. Figure 3.70 shows the same pattern after the background has been subtracted and the peaks have been fitted. The difference pattern is at the bottom. Details can be found in Hüpen et al. (1991).

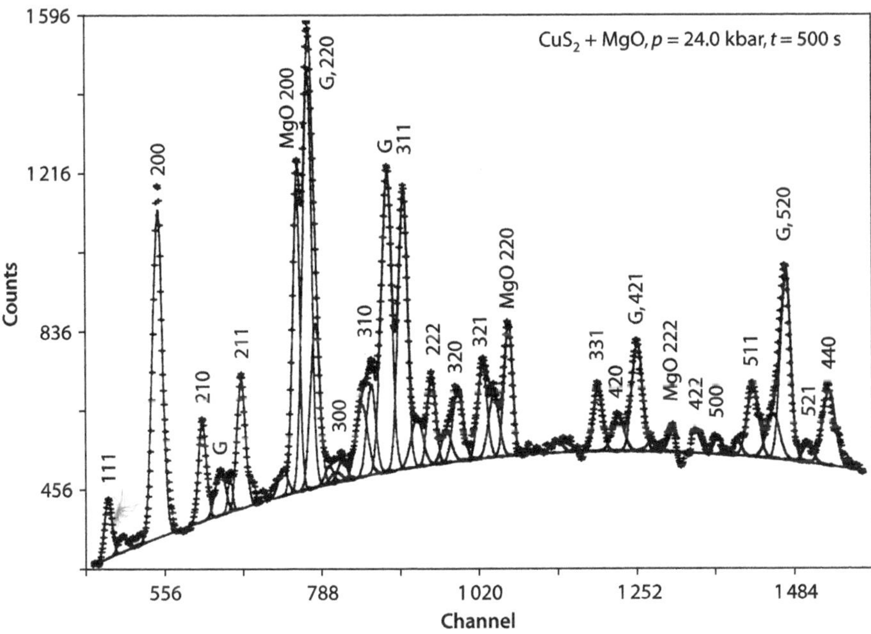

Fig. 3.68. Diffraction pattern of CuS₂ + MgO (pressure marker) collected at $p = 24$ kbar

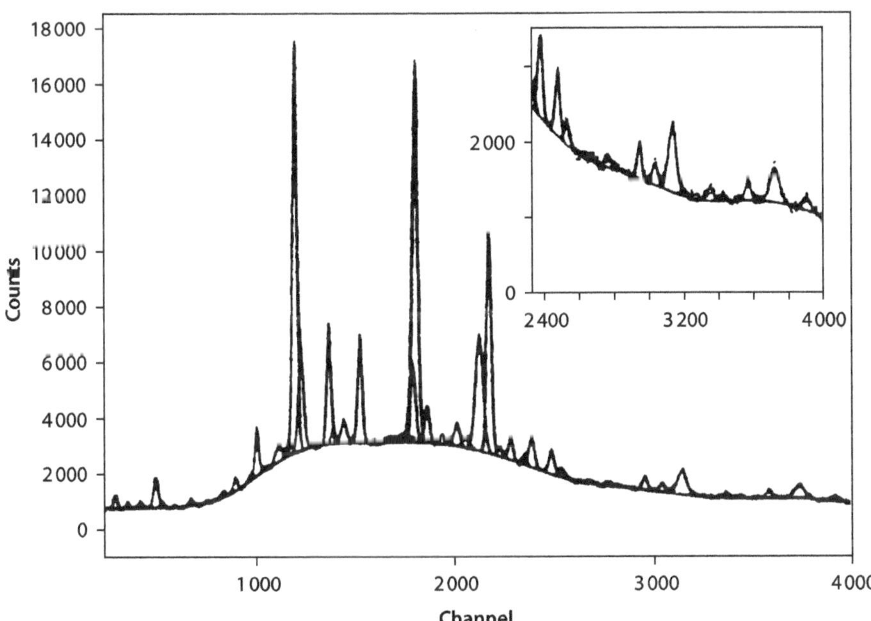

Fig. 3.69. Raw diffraction pattern of CuS₂ + MgO as measured prior to analysis

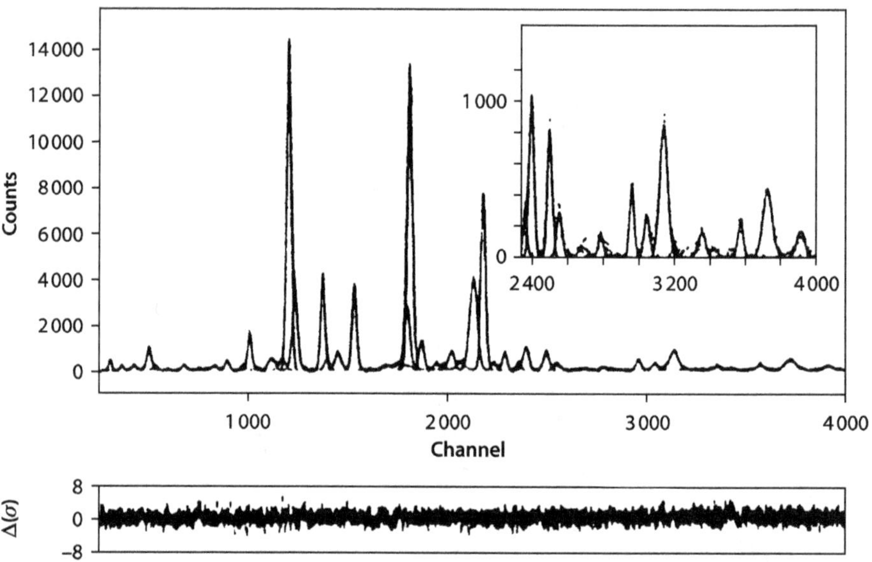

Fig. 3.70. Diffraction pattern of Fig. 3.68 after analysis: background has been subtracted, peaks have been fitted

A number of pyrite type compounds with 3d elements have been studied in this project (Will et al. 1984). In general the compressibility was the aim of the study, but some compounds including, CuS_2 experience also a phase transformation at elevated pressures from the Fe-pyrite type structure to the marcasite structure. In this specific example of copper pyrite there is a special and specific problem, since the parameters of the two unit cells are very close to each other: cubic CuS_2-pyrite with $a_0 = 5.790(1)$ Å, orthorhombic CuS_2-marcasite with $a_0 = 4.70(1)$ Å, $b_0 = 5.80(1)$ Å, $c_0 = 3.54(1)$ Å (extrapolated to zero pressure). The two phases therefore have diffraction lines very close to each other. Separation is only possible by interactive profile fitting.

3.10.5
Cation Distribution in a Thin Film Garnet Sample

The structural characterization of a garnet thin film sample is an especially interesting case where the Two Stage method proved to be very successful and basically the only method to give the wanted information. The sample discussed here was a magnetic garnet film of composition $(BiY)_3(FeGa)_5O_{12}$ deposited in Argon atmosphere onto a glass substrate. As-deposited films were found by X-ray diffraction to be amorphous. Annealing at 650 °C for three hours lead to a crystalline diffraction pattern. With this specimen data for the structural analysis were collected with a Siemens D5000 diffractometer using a Rigaku rotating anode target. The diffraction diagram of a 3 400 Å thick specimen is depicted in Fig. 3.71. It is discussed in more detail. Step width was $\Delta(2\Theta) = 0.02°$ with 30 seconds per step. The insert in Fig. 3.71 shows the full diagram

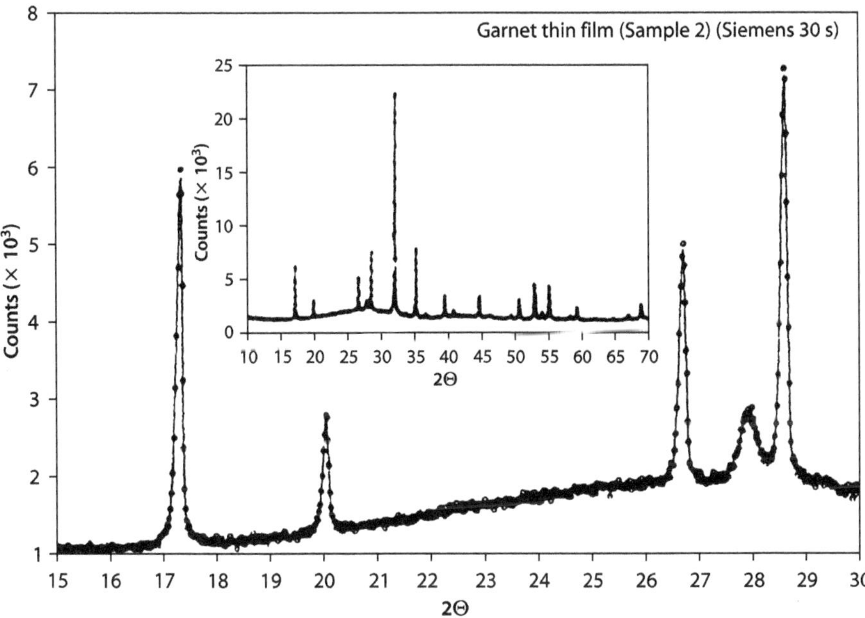

Fig. 3.71. X-ray diffraction pattern of a 3 400 Å thick polycrystalline garnet sample. The *insert* shows the full diagram from 10 to 70°

from 10 to 70° with the strongest peak 422 with an intensity of 22 000 counts. In both patterns the background coming from the amorphous glass substrate is clearly visible.

The analysis followed the routine of the Two Stage method: Subtracting background, followed by profile fitting. This yielded peak positions, integrated intensities and peak full width at half maximum, *FWHM*. From the peak positions a lattice parameter $a = 12.512(4)$ Å was calculated.

Figures 3.72 and 3.73 show two section of the diffraction pattern, one at low angles and high intensity, the other at high angles and low intensities. The profile fitting together with the differences at the bottom is included.

Using the program POWLS a structural model must be fitted to the observed intensities. Garnets with the general formula $A_3B'_2B''_3O_{12}$ crystallize with cubic symmetry in Ia3d with A^{2+} in 24c (1/8,0,1/4); B'^{2+} in 16a (1/4,1/4,1/4), B''^{4+} in 24d (3/8,0,1/4) and oxygen in 96h (x,y,z). In mineral garnets B" is silicon. With the given analysis the composition of the specimen is.

$$(BiY)_3(FeGa)_5O_{12}$$

We have four cation species to be distributed over three cation sites. With only one experimental data set, e.g. without for example neutron diffraction data or anomalous X-ray dispersion data an unique solution is not possible. A possible approach can be followed exhausting the potentials of POWLS.

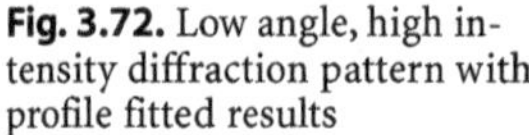

Fig. 3.72. Low angle, high intensity diffraction pattern with profile fitted results

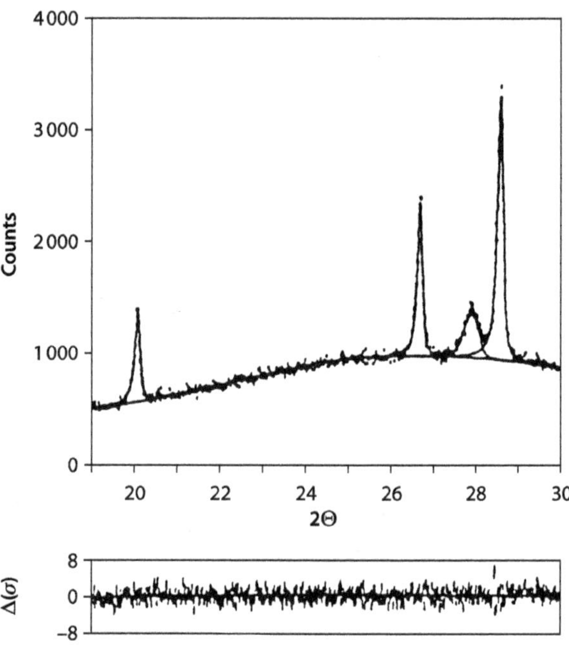

For the cations in the sample: Bi, Y, Fe and Ga, the shape of the form factors is very similar. We can scale them to $f = 1.0$ at $(\sin\Theta / \lambda) = 0$. Deviations including the ionisation states are found only in the low angle region about at $(\sin\Theta / \lambda) < 0.18$ Å^{-1}. The analysis is done therefore only on the occupation numbers of each cation species OC_j. For oxygen we set $OC = 12$ as an internal standard. Additional parameters are x, y, z for oxygen and the (isotropic) temperature factors B_j for all atoms. The refinement with POWLS was done in alternating blocks:

- x, y, z, K (scale factor), and B_j
- the OC_j numbers

Within a few cycles R-values around 3.5% were reached. The POWLS least squares calculations yield the numbers of electrons in each Wyckoff position. This is shown in Table 3.23 in Column 2. As a next step these numbers have to be brought into agreement with the analysis, e.g. the cations in the structure. Referring to a reference compilation of garnets made by Geller (Geller 1967), Bi can only go to 24c. Since there are only 2.5 Bi ions, some of the iron ions must also got into 24c statistically with Bi. Depending on the ionisation state Fe has been found in any of these sites in previous investigations (Geller 1967). The site 24d, occupied by Si^{4+} in minerals, was assumed to contain all gallium ions, 1.0 Ga, plus 2.0 ions of Fe^{4+}. This leaves 16a to be filled with 1.0 Y (according to Geller it can be found in 16a or 24d) plus 1.0 Fe. The resulting formula, with the final values included in Table 3.23 is then:

$$(Bi_{2.5}Fe_{0.5})(YFe)(GaFe_2)O_{12}$$

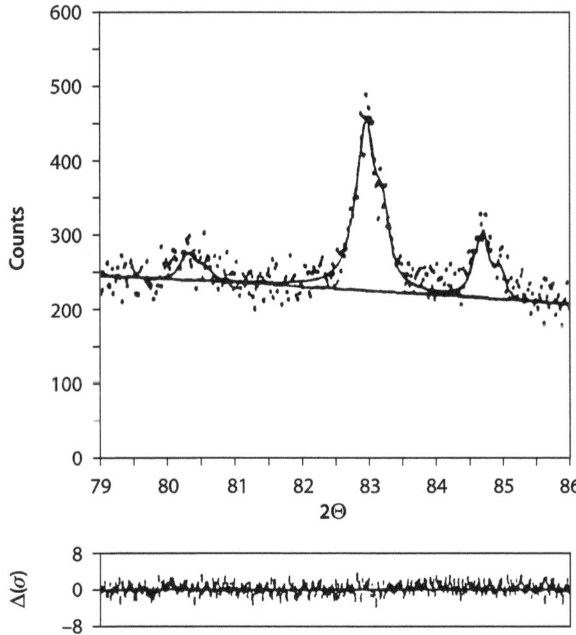

Fig. 3.73. High angle low intensity diffraction pattern with profile fitted results

Table 3.23. Final results of the least squares refinement with POWLS of the garnet data

Wyckoff site	Number of electrons calculated by POWLS	Occupation by cations	Number of electrons of cations
24 c	196	2.5 Bi + 0.5 Fe	211
16 a	54	1.0 Y + 1.0 Fe	59
24 d	73	1.0 Ga + 2.0 Fe	71

The POWLS calculation requires the number of electrons in each site as listed in Table 3.23, a total of 323 electrons. Accepting the distribution discussed the final result gave the numbers shown in the last column of Table 3.23, a total of 341 electrons. The agreement is satisfactory. Discrepancies are found for Bi, with a possible explanation coming from uncertainties in the analysis. This was based on micro probe analysis and X-ray fluorescence. The strong variation of the *FWHM* data, depicted in Fig. 3.75, suggests a non-uniform composition and this may explain our findings. The full analysis is consistent in itself.

Analysis of the FWHM

The reflections in the diffraction pattern are broadened in comparison with a standard silicon specimen (Figs. 3.72 and 3.73). In a routine analysis as reported here the *FWHM* is always determined prior to the actual specimen measurement using a sili-

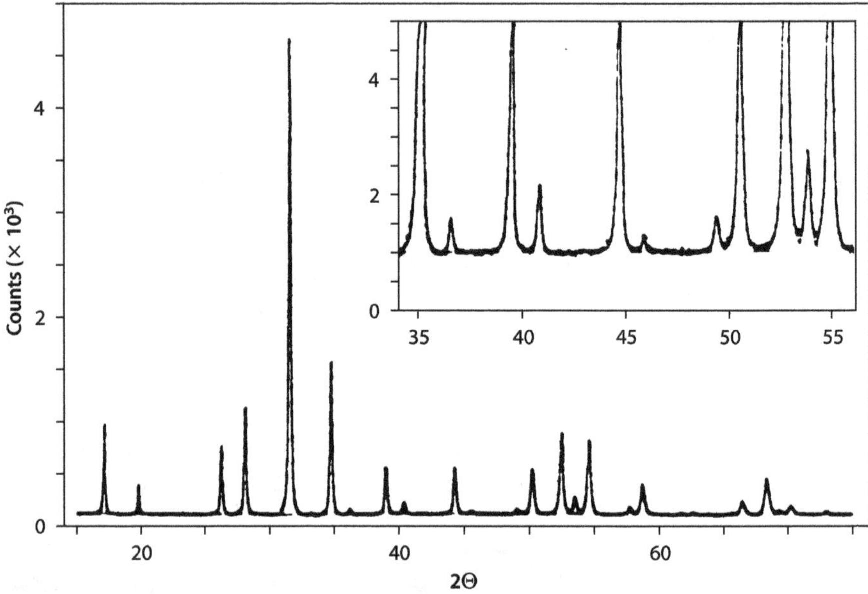

Fig. 3.74. The X-ray diffraction pattern after background subtraction. The *insert* gives a selected part showing the profile fitted data to separate overlapping peaks

Fig. 3.75. Angular dependence of the *FWHM* for a garnet sample (*upper curve*) in comparison with a silicon sample (*lower curve*). A total of three measurements were analyzed, plotted as triangles, squares and diamonds

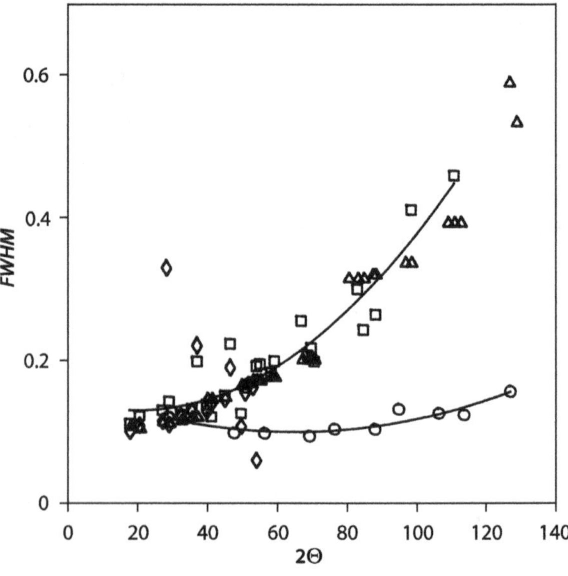

con standard specimen. In this specific case the profile fitting procedure gave the values as plotted in Fig. 3.75. When the *FWHM* was plotted versus $1/\cos\Theta$, by using the

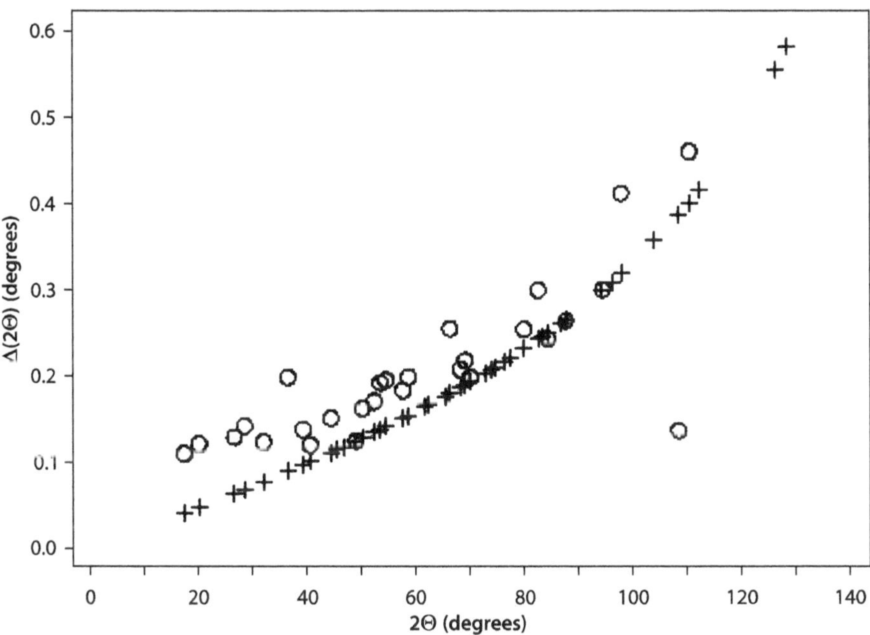

Fig. 3.76. Comparison of measured *FWHM* with *FWHM* values calculated for a specimen with a variation of lattice parameters by ±0.01 Å

Warren-Averbach method (Warren and Averbach 1950, 1952), a particle size of about 100 Å was calculated, in agreement with electron microscopy analysis. Another and final explanation is found by assuming deviations in the stoichiometry of the individual particles in the sample. If we assume a variation within the specimen of $\Delta a_0 = \pm 0.01$ Å we obtain a curve depicted in Fig. 3.76, which produces the measured *FWHM* data perfectly. That means the (broadened) peaks are actually a superposition of many narrow peaks of slightly different unit cell sizes clustered together. With increasing 2Θ these peaks are spread more and more. Also it is not unreasonable to assume that the lattice parameters vary due to a variation of the stoichiometry, if we remember how the sample was prepared. In conclusion the analysis of this rather unconventional specimen was very successful.

3.11
Structure Determination from Energy Dispersive Data

3.11.1
Structure Determination from Energy Dispersive Data

The generally and routinely applied method for data collection is by scanning the specimen with a detector as a function of angle 2Θ, the angle dispersive mode. Also position sensitive detectors, which collect data simultaneously at a number of scattering angles 2Θ, operate in the angle dispersive mode. An alternative to 2Θ scanning is the recording of diffraction data as a function of energy, the energy dispersive mode. Energy E thereby replaces the scattering angle 2Θ on the abscissa. The transformation from scattering angles to energy is straight forward:

$$2d_H \sin\Theta_0 = \lambda_H = \frac{hc}{E_H} = 12.4\,(\text{keV}\cdot\text{Å})\frac{1}{E_H} \tag{3.46}$$

H stands for the Miller indices hkl and $2\Theta_0$ for the fixed scattering angle. Energy dispersive diffraction has emerged from making use of the continuous primary radiation of high intensity and high energies of synchrotrons. But also in neutron diffraction energy dispersive measurements have recently become an important method when using spallation sources.

Not so simple is the analysis of the diffraction patterns, mainly because the profiles of the diffraction peaks are more difficult to describe and more over are often dependent on energy. As a consequence the software for analysis has stayed behind. The programs for total pattern fitting presented by Pawley (1980, 1981) and Jansen E. et al. (1988) are written for the analysis of conventional, angle dispersive diffraction patterns, and there are no programs generally available for full pattern analysis in the sense of Rietveld refinement.

3.11.2
Synchrotron Radiation

Energy sensitive solid state detectors, as they are used in diffraction experiments with synchrotron radiation, lack the resolution compared to proportional or scintillation single channel counters. Therefore this technique is used mostly for measurement with the specimens under extreme condition, under high pressure, high or low temperatures, or for time dependent experiments, for example to study phase transformations (Will and Lauterjung 1987; Will and Berndt 2001). For such investigations energy dispersive powder diffraction offers a number of advantages because the complete diffraction pattern is recorded at any instance, and it can be recorded extremely fast. This weighs against the disadvantages: Low resolution, unavoidable escape peaks, or "impurity peaks" in patterns coming for example from the pressure marker when collected with the specimen under high pressure. Another serious difficulty comes from the high and with energy variable background. These deficiencies render the conventional analysis techniques almost hopeless except for very simple crystals. Therefore such patterns are usually analyzed in a more elaborate way by "interactive pattern decomposition" (see Sect. 3.3). Consequently little attention has been paid to full pattern analysis of energy dispersive diffraction patterns for structural purposes.

There are a few successful attempts to analyze energy dispersive X-ray patterns. One such program for total pattern fitting of energy dispersive powder diffraction pattern has been presented by Honkimäki and Suortti (1992). They followed the procedure of decomposition the pattern into its individual Bragg peaks for further analysis of crystallite size and strain. In their example the radiation source was a tungsten X-ray tube, emitting a continuous energy spectrum, but their method can be applied equally well for patterns collected with synchrotron radiation. The total observed pattern is taken as a sum of incoherent and coherent scattering. The incoherent part leads to a considerable contribution of a very high background. This can be seen in the figures presented in their paper; and see also Fig. 3.68. The background is coming from incoherent and coherent scattering. The incoherent part has been calculated from theoretical cross sections for the individual atoms. The coherent part is described as a sum of discrete Bragg peaks and acoustic and optic-phonon thermal diffuse scattering, TDS. This is calculated from Debye and Einstein models, respectively. The model pattern is convoluted by the instrument function, and the total pattern is fitted to the observed one by varying the integrated Bragg intensities.

The method has been applied to patterns of Mg, Al and Ti, compounds with very simple structures and where data are available for calculating coherent and incoherent contributions. In conclusion, the analysis gave very good results, however the procedure for fitting is tedious and it must be doubted whether it is widely applicable.

3.11.3
Neutron Diffraction by Time-of-Flight Measurements

"Energy dispersion" diffraction with neutrons is conducted by time-of-flight techniques, TOF. This requires a "white" neutron beam and an appropriate analysis of these data. This has been realized in the beginning with a chopper at steady state reactors. A chopper produces a pulsed "white" neutron beam, and the diffraction patterns from a specimen are obtained with time-focused counter arrays, BF_3 or helium filled. The peak shapes in such a setup are generally Gaussian with nearly constant resolution. In these cases Rietveld refinement technique can, and has occasionally been successfully used to obtain structural and profile parameters. Worlton et al. (1976) investigated for example KCN IV and Si_3N_4 under pressure as an example using TOF neutron data They analyzed their data by a multicomponent profile refinement because of spurious peaks from foreign materials in the sample. These authors "considered Rietveld analysis hopeless, except for the simplest crystals". They analyzed their patterns of the multicomponent sample, four phases, by a decomposition process into separate peaks.

Alternatively, we have seen recently the development of accelerator based pulsed neutron sources, specifically at the Argonne National Laboratory, Argonne, Ill., USA, and at ISIS at the Rutherford Appleton Laboratory in Chilton, UK. This has led to wide spread use of time-of-flight technique for neutron powder diffraction. In spallation sources neutron are generated by the interaction between a beam of high energy protons or electrons and a heavy metal target, usually tungsten or depleted uranium. The time between the pulsed source at ISIS is 20 ms, much greater than the duration of the burst itself of only a few µs. On striking the target each proton produces 25–30 fast neutrons by the spallation ("nucleus chipping") process. These neutrons are high energy neutrons, which have to be slowed down, e.g. moderated to thermal and epithermal energies before they are used for diffraction. Diffraction at pulsed spallation neutron sources have overcome the shortcomings of TOF diffraction on reactor sources. The moderation process and the geometrical contributions by the diffractometer and detector leads to a non-Gaussian and distinctly asymmetric diffraction line profile. The peak shapes and its wavelength dependence are the short-comings. This makes full pattern or Rietveld analysis difficult, or it requires extensive programming (see e.g. Lager et al. 1981).

Von Dreele et al. (1982) have adapted a much modified program of the FORTRAN program originally written by Rietveld (1969) for the analysis and refinement of TOF data. The computation is done in two steps, consisting of data preparation and a least squares step. The useful application of this computer code has been demonstrated by refinements of Al_2O_3 (Jorgensen and Rotella 1982) or of forsterite (Lager et al. 1981). In case of forsterite neutron powder data were collected in backscattering geometry by a detector bank at the high resolution time-of flight powder diffractometer station at Argonne National Laboratory's ZING-P' pulsed neutron source. Each detector array, at the right and left side, contained 16 ^{3}He counters. The data were refined using Von Dreele's program modified for TOF data from spallation pulsed neutron sources. The asymmetric non-Gaussian peaks were fitted with six parameters. The ensuing Rietveld refinement included 2 519 data points from 464 allowed reflections. Forty seven parameters were refined. The final R-value was 2.10% including background, and 3.53% after the background had been removed. The agreement between the results obtained from TOF data and the results of single crystal data analyses was very good.

The advantages of TOF compensate for the disadvantages. The properties of the pulsed neutron beam make it possible to achieve unusually high resolution over a wide range of d-spacings, from $d = 0.5$ Å to more than 5.0 Å, high count rates and the ability to collect a complete data set at fixed scattering angles. It is specifically the speed and simplicity of data collection which make TOF measurements so attractive.

Neutrons are relatively slow particles. The transformation from energy data to 2Θ data is easy and straight forward. If p is the total flight path, moderator-sample-detector, we have $p = vt$, with t the time of flight in seconds and v the neutron velocity. In the TOF method the wavelength λ is determined by the flight path length p (in m) and flight time t through the de Broglie relation

$$\lambda = \frac{h}{mv} = \frac{h}{m}\frac{t}{p} = 3.96 \times 10^3 \frac{t}{p} \tag{3.47}$$

3.11.4
Simultaneous Angle and Position Resolving Neutron Diffraction by Time-of-Flight Measurements

Recent developments in position resolving detectors have led to a completely novel experimental situation at spallation sources. The instrumental setup is simple, if a position sensitive detectors, for example JULIOS, is used (Will et al. 1994). One such setup has been installed at the ROTAX station at ISIS at the Rutherford Appleton Laboratory in Chilton, UK. At ISIS the neutrons come in bursts of a few μ-seconds. Unlike in other detection setups, they hit the position resolving detector, e.g. the diffracted radiation is recorded simultaneously at different angles. The range of wavelength which can be recorded depends on the scattering angle 2Θ, at which the detector is set. Figure 3.77 gives schematically three standard positions for the detector, in forward direction, at 2$\Theta = 90°$, and in a backscattering position. This allows the recording of

d-values between 2 and 35 Å (forward diffraction), 0.6 to 4 Å (90°) and 0.3 to 1.2 Å (backscattering). Figure 3.78 shows schematically the accessible d-ranges for the JULIOS units in three standard settings in terms of wavelength λ versus scattering angle. The shaded boxes represent the areas available with the shown three detector settings. The solid lines represent selected bands of constant d-spacings in Å, which are characteristic if the unit is centered at a (fixed) 2Θ angle of $2\Theta = 23°$ (forward direction), at $2\Theta = 90°$ or at $2\Theta = 137°$ in backscattering position. The specimen is always positioned directly in the white beam. With the three settings shown the wavelength band available to the user is approximately form 0.3 Å theoretically to 50 Å.

JULIOS is a linear position sensitive scintillation detector constructed for simultaneous recording of time-of-flight and angular dispersive data. A detailed descriptions can be found in papers by Schäfer et al. (1992, 1995). The pulsed-source TOF experi-

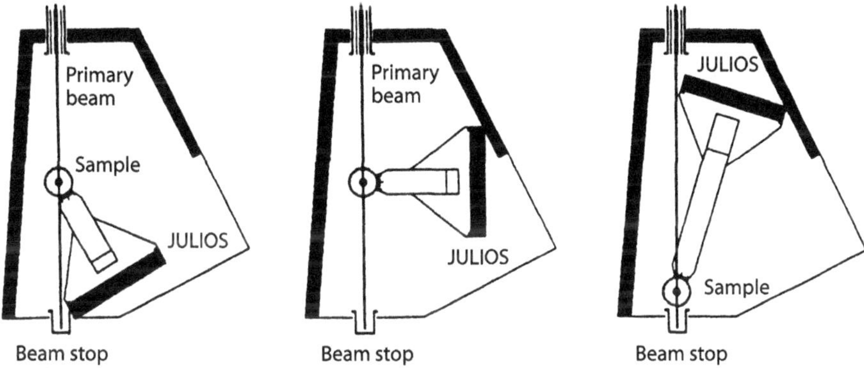

Fig. 3.77. Experimental arrangement of the JULIOS detector in three different positions: forward, 90° and backscattering

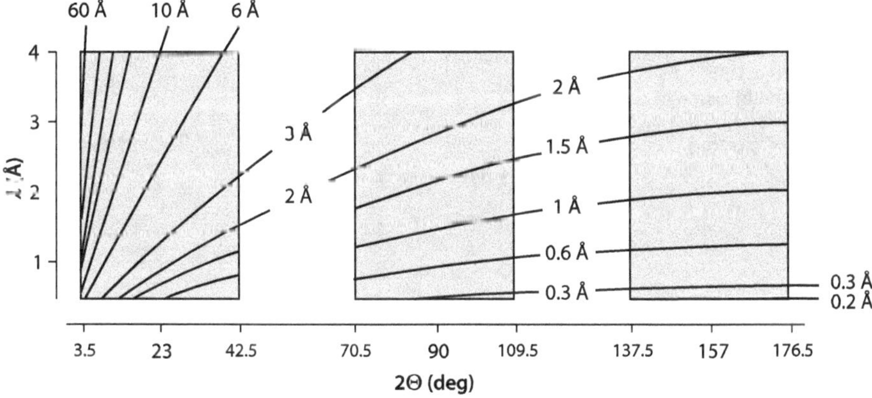

Fig. 3.78. Schematics of parameter sets λ vs. scattering angle 2Θ available at three different standard settings as indicated in Fig. 3.77: forward, 90° and backward. The detector units are centered at respectively $2\Theta = 23°$, 90° or 157°. The *shaded boxes* indicate the data range which can be recorded

ments are based on the simultaneous time- and position resolving scintillation detector. The detection zone in this detector is only 1 mm deep and at the same time is highly sensitive over the entire thermal energy spectrum. This makes this detector especially well suited for applications in TOF technology at pulsed spallation sources. Before installation at ISIS it had been in routine operation at reactor crystal-monochromator diffractometers. The great improvement of this simultaneous angle and position resolving neutron diffraction by TOF measurements can be seen by comparing in Fig. 3.79 conventional diffractometry at a steady state reactor and a white beam diffractometer at a pulsed source. The utilization of the primary thermal neutron spectra in both configurations is shown by the hatched sectors $\Delta\lambda$, a small monochromatic band at the reactor, and a wide energy area when using white beam neutrons.

The measurements result in a two parameter diagram of scattering angle 2Θ, the channel in the detector, and wavelength λ. A typical two-parameter pattern of an angle dispersive TOF measurement registered with the JULIOS scintillation detector at ISIS is shown in Fig. 3.80a for polycrystalline corundum, Al_2O_3. The *hkl* reflections are observed as intensity bands (compare Fig. 3.78), which are registered in a set of $(2\Theta$-$\lambda)$ detector elements: identical *d*-spacings are seen by a multitude of different detector elements. The count rates of equal *d*-channels are accumulated afterwards into a one dimensional diagram during the analysis and normalized for different channel occupation. This is shown in Fig. 3.81.

Figure 3.81 shows the one-dimensional diffraction pattern of corundum corresponding to Fig. 3.80a. The wavelength band was from 0.9 to 4.7 Å, the detector covered a 2Θ range from 74.8 to 107.0°.

Alternatively the parameters in the *xy*-plane can be used as time (channel) vs. position (channel). Such a pattern is shown in Fig. 3.80b for nickel. The transformation from one scale to another is straight forward. In the case of Ni shown here the angular 2Θ range was 60 to 90°, and the wavelength range $\lambda = 1.0$ to 2.8 Å, corresponding to

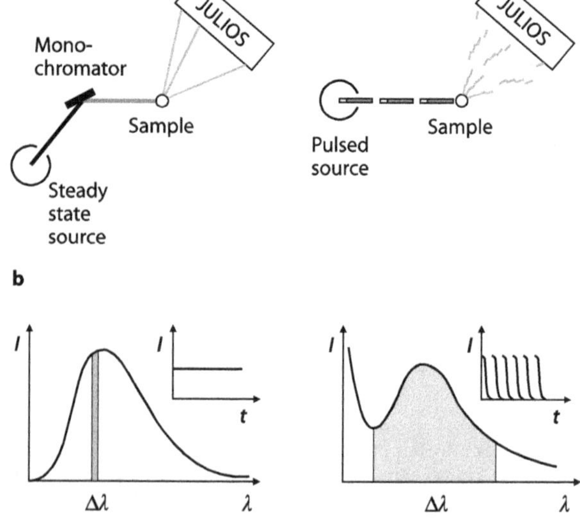

Fig. 3.79. Conventional diffractometry at a steady state source (*left*) and a white beam diffractometer at a pulsed source (*right*); **a** schemes of experimental setups with JULIOS detectors; **b** utilization of the primary thermal neutron spectra (*hatched sectors* $\Delta\lambda$) in both configurations

3 to 9 ms. One channel corresponds to 2.3 µs, and on the other channel to 0.12°. The exposure time was 1 hour. The background is very low, with 3 counts per hour per chan-

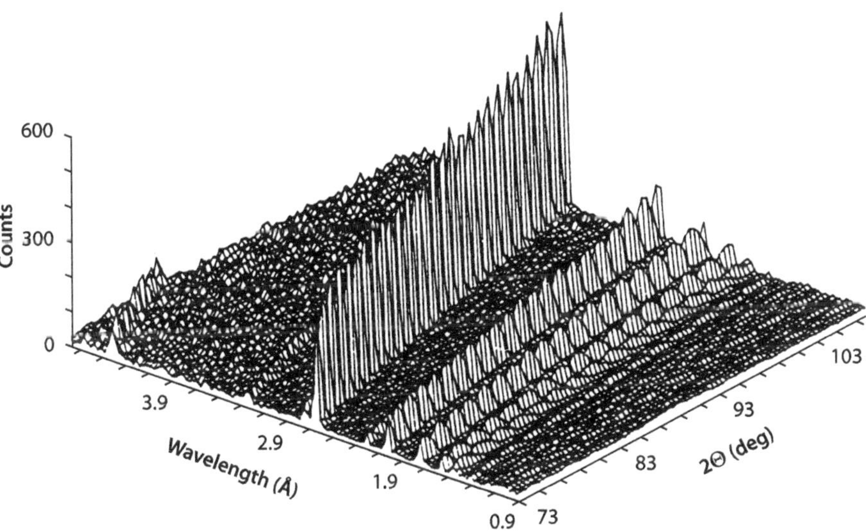

Fig. 3.80 a. Typical two parameter pattern, 2Θ vs. λ, showing reflection bands, here for corundum, collected with the JULIOS detector in angular dispersive TOF technology

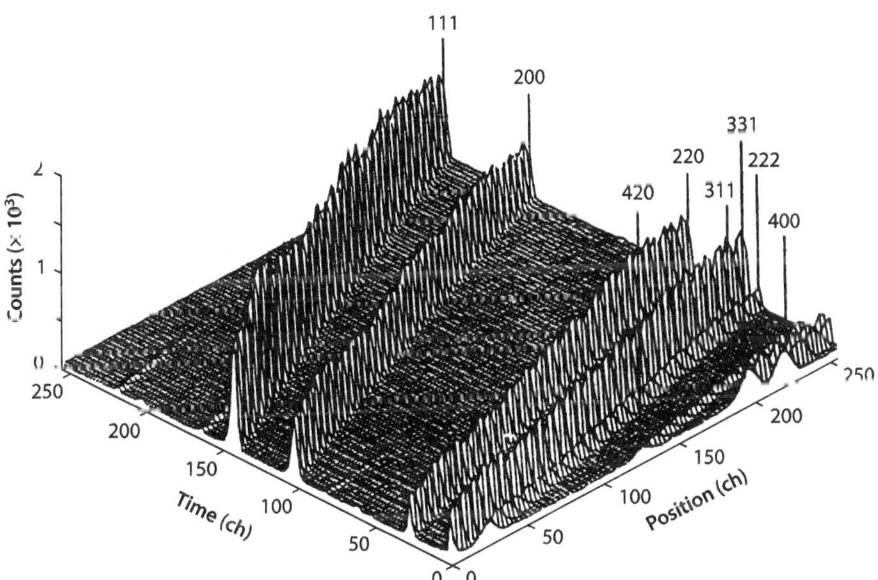

Fig. 3.80 b. Two parameter pattern for nickel with the parameters position (in the detector) versus time, showing the typical reflection bands

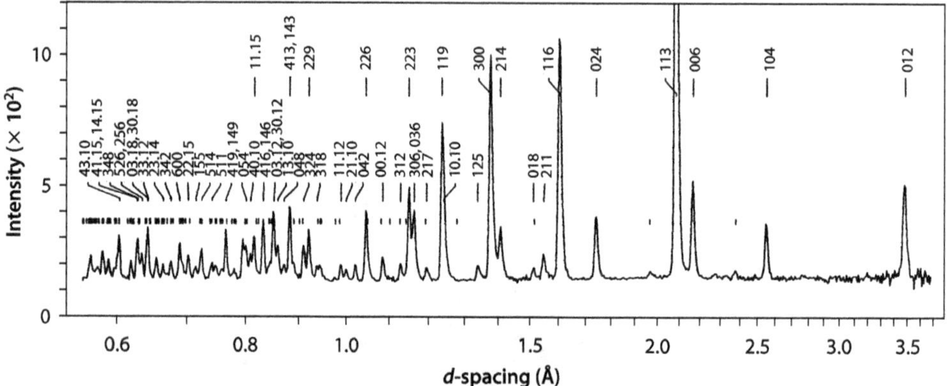

Fig. 3.81. One-dimensional d-spacing pattern of corundum calculated from the data of Fig. 3.80a

nel giving a ratio peak:background = 30:1. The translation into a one-dimensional diffraction pattern is included in Fig. 3.82.

ISIS is operated at a 50 Hz repetition rate, corresponding to a 20 ms neutron frame. The d-spacing limits of the instrument are determined by the limiting $(\lambda/2\Theta)$ parameter settings covering normally d-values from 0.2 to 20 Å. Figure 3.82 depicts four examples of typical TOF diffraction patterns collected with JULIOS at different conditions with d-spacing ranging from 0.3 to 15 Å. NiO and Nickel have been measured in backscattering geometry, corundum in a 90° detector setting and mica in forward scattering geometry.

3.11.5
Electronic Zooming

An additional advantage of this technique and setup is the possibility of "electronic zooming", e.g. the possibilities to vary and improve the resolution by adjusting the d-scale to wanted experimental requirements by spreading independently either the position or the time window. This is called "electronic zooming". It can be position- or time zooming. This improvement is found only in combination of TOF with position resolving scintillation detectors. The width of the d-channels is preselected by software at the beginning of the measurement when the time and position windows of the detector are set. By setting different, e.g. smaller windows the resolution can be changed and improved. Figure 3.83 gives an example of zooming the position scale of the detector for a quartz measurement. The spectrum (a) at the top is recorded with a time window from 3.0 to 9.0 ms. The maximum number of the 256 position channels of the detector are at the top picture distributed over an area of 22 activated photomultipliers covering a 2Θ section of 31° and a d range from 0.9 to 6.4 Å; the channel width in 2Θ is 0.12°. In the lower diagram the spectrum is distributed by zooming over only 8 adjacent photomultipliers covering a 2Θ section of only 12° with a d range from 1.2 to 4.6 Å, but with the channel width reduced to 0.04°. The pattern is recorded with

Fig. 3.82. Four typical TOF diffraction patterns obtained with one single JULIOS unit: NiO and Nickel in backscattering geometry, corundum in a 90° detector setting and mica in forward scattering geometry. The time to collect one pattern was 1 hour. The range of d-spacings obtained is clearly visible

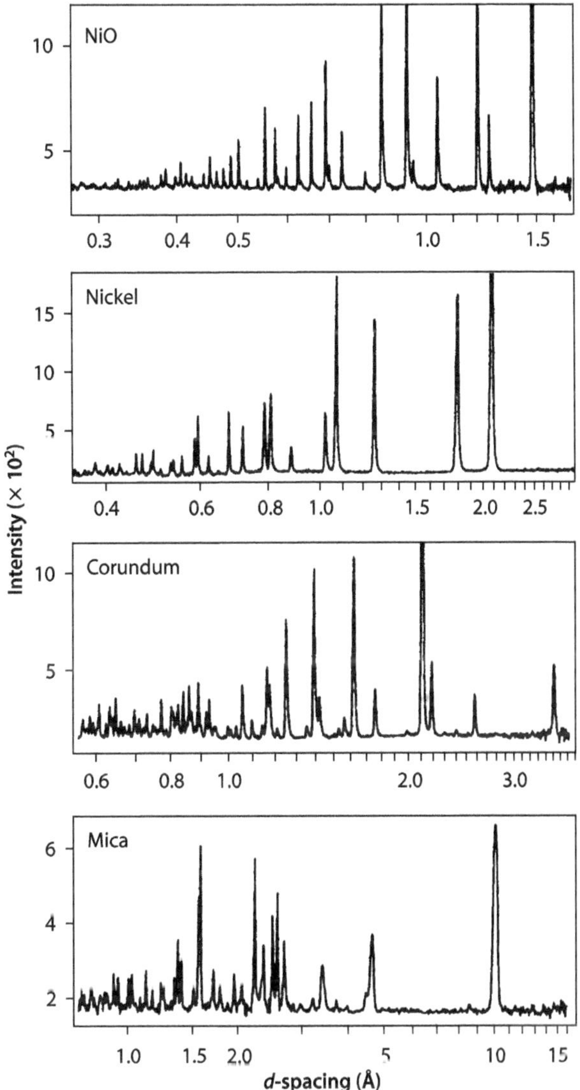

still 256 position channels. Both spectra were recorded at the same fixed detector position of $2\Theta_{center} = 40°$.

Similarly we can "zoom" the time scale, as is shown in Fig. 3.84 for the same quartz specimen. The detector is centered at $2\Theta = 70°$. At the top the maximum number of 256 time channels was distributed over a time window from 3.0 to 7.5 ms. In the lower diagram the spectrum is distributed by zooming over 16 activated photomultipliers and the time window is reduced to 3.0 to 5.25 ms with a d-range from 0.7 to 1.6 Å. The pattern is

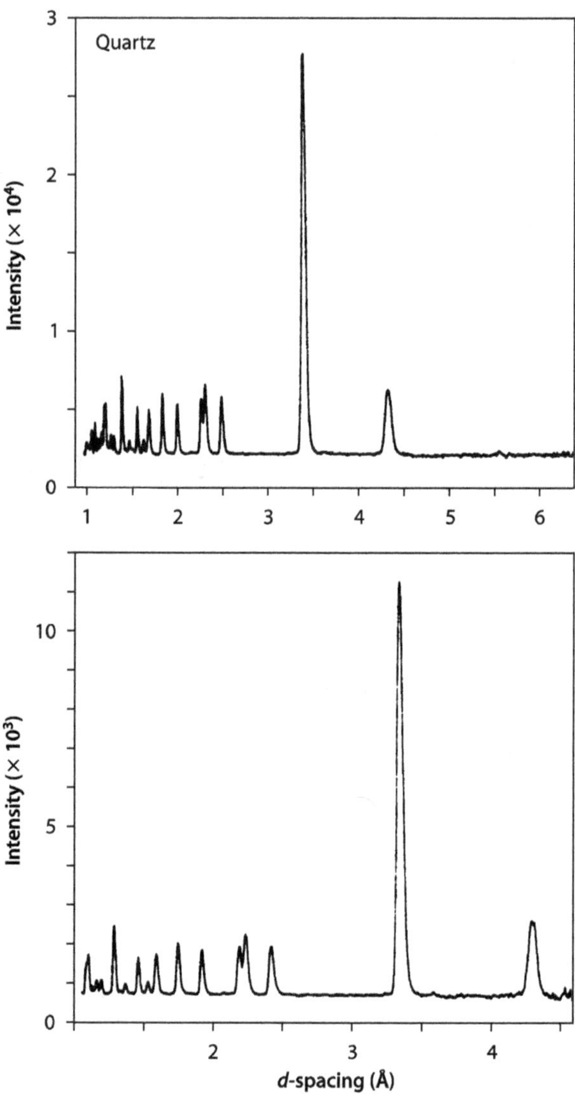

Fig. 3.83. Quartz spectra obtained by zooming the position scale of the detector; at the *top* the full pattern is shown, the *lower picture* shows a reduced section with higher resolution by "zooming"

still recorded with 256 position channels. The channel width $\Delta\lambda$ is reduced from 0.0054 to 0.0027 Å. Both spectra were recorded at the same fixed detector position of $2\Theta_{center} = 40°$.

3.11.6
Analysis of TOF Data

Since data are recorded in a 3-dimensional two parameter pattern, there are no peaks in the usual sense. The diffraction patterns collected by this method are intensity bands

Fig. 3.84. Quartz spectra obtained by zooming the time scale of the detector; at the *top* the full pattern is shown, the *lower picture* shows a reduced section with higher resolution by "zooming"

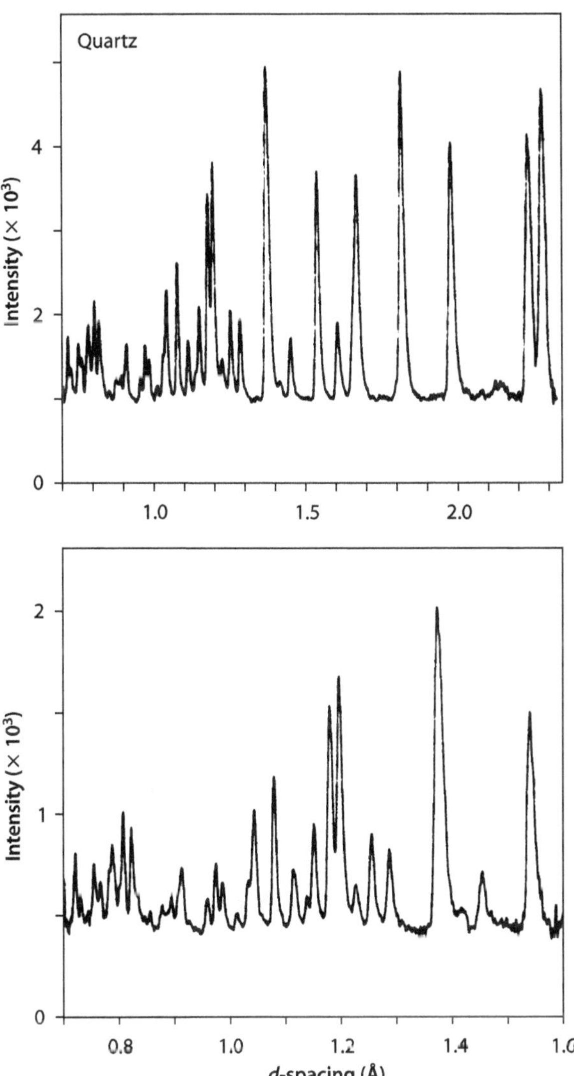

at the 2Θ-wavelength scale (see Fig. 3.80). The moderation process and the geometrical contribution by the diffractometer yields a non-Gaussian and distinctly asymmetric diffraction line profile. The analysis follows therefore the Two Stage method. Also the calculation of intensities in the least squares routines, for example POWLS, require special attention.

Angle dispersive TOF diffraction patterns are composed of a set of Debye-Scherrer patterns registered in position channels. For the analysis they are recalculated fore equidistant $\Delta\Theta$, and of a set of TOF patterns collected in constant $\Delta\lambda$ channels

(Fig. 3.80). Bragg intensities P_Θ of Debye-Scherrer patterns from a cylindrical specimen, registered in constant 2Θ steps, are defined by

$$P_\Theta \sim I_0(N^2hV / 8\pi r)(\lambda^3 / \sin\Theta \sin2\Theta)jF^2 \tag{3.48}$$

with I_0 the primary intensity, N the number of unit cells per unit volume, V the effective specimen volume, h the detector height and r the detector distance, j the multiplicity factor and $F = F_{hkl}$ the structure factor. The Bragg intensities P_λ of TOF patterns of constant λ steps are defined by

$$P_\lambda \sim I_0(N^2hV / 16\pi r)(\lambda^4 / \sin^3 \Theta)jF^2 \tag{3.49}$$

For details see Bacon (1975). The conversion of both $\Delta\Theta$ and $\Delta\lambda$ patterns into d-spacing intensity patterns P_d, linear in d-spacing, is a result of Bragg's law; partial derivatives result in

$$\Delta d / \Delta\Theta \sim d(\cos\Theta / \sin\Theta) \tag{3.50}$$

and

$$\Delta d / \Delta\lambda \sim 1 / (2\sin\Theta) \tag{3.51}$$

With Eqs. 3.50 and 3.51 and the trigonometric formula $\sin2\Theta = 2\sin\Theta \cos\Theta$, it follows

$$P_d = P_\Theta(\Delta d / \Delta\Theta) = P_\lambda(\Delta d / \Delta\lambda) \tag{3.52}$$

and finally

$$P_d \sim I_0(N^2hV / 2\pi r)d^4jF^2 \tag{3.53}$$

Owing to the nonlinear experimental channel density in d-space and in view of a more favourable representation of d-spacing patterns, it is convenient to perform a recalculation of intensities $P_{\ln(d)}$ which are linear in the logarithm of d with

$$\Delta d / \Delta\ln(d) = 1 / d \tag{3.54}$$

The corresponding intensity formula is then

$$P_{\ln(d)} = P_d(1 / d) \sim I_0(N^2hV / 2\pi r)d^3jF^2 \tag{3.55}$$

With this intensity formula integrated intensities can be extracted form the $2\Theta/\lambda$ diffraction pattern and used for further analysis in case of unknown crystal structures, or of magnetic structures, or for the refinement with POWLS.

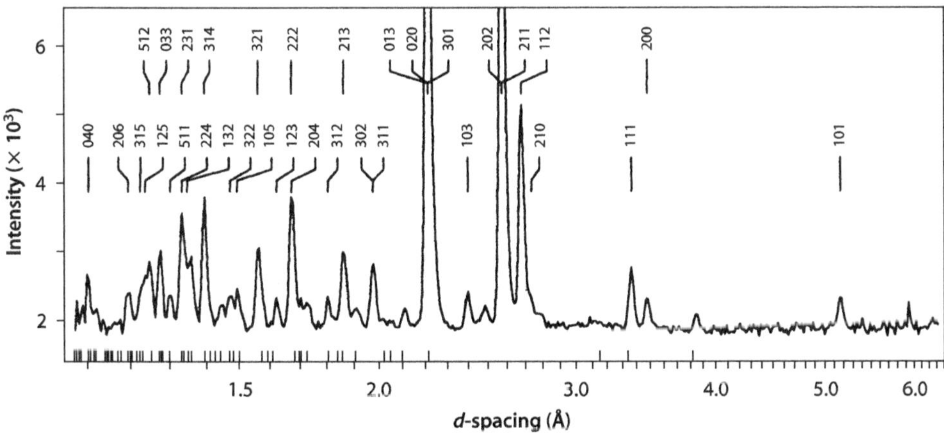

Fig. 3.85. Neutron diffraction d-spacing pattern of TbPtGa taken at 300 K

A considerable number of diffraction patterns have been measured and analyzed. Analysis and refinement was done according to the Two Stage method. With integrated intensities or groups of intensities the structures can be refined, for example with POWLS. As just one example we mention the determination of the crystallographic and magnetic structure of TbPtGa (Schäfer et al. 1994). The diffraction d-spacing pattern taken at 300 K is shown in Fig. 3.85.

3.11.7
Laue Diffraction with Neutrons by Time-of-Flight Measurements

A very promising application of a position sensitive detector at a pulsed neutron beam is a completely new way to study single crystals by Laue diffraction. This is not with powders, but it is worth mentioning it briefly in this monograph.

This application is possible and very easy with an experimental setup like it is at ISIS: a pulsed white neutron beam and the position sensitive detector JULIOS. (Will et al. 1994). Figure 3.86 shows a three dimensional picture of a Laue diffraction pattern of a Ge single crystal. It is well known that Laue peaks, for example in X-ray diffraction experiments, contain superimposed the higher orders of the same reflection, for example n(111). TOF measurements in a pulsed beam separate the higher orders and therefore allow one to perform single crystal measurements easily. The picture shows the higher orders nicely separated. The Ge crystal had a distance to the detector of 37 cm. The sensitive detector window covered a 2Θ region of about 60 to 125°. The angular range was λ=0.5 to 2.5 Å, corresponding to a time window of $\Delta t = 2$ to 8 ms. Fourteen reflections have been measured simultaneously with a maximum count rate of 40 000 for the (400) reflection.

Here the circle closes to the very first experiments by Laue, Friedrich and Knipping in 1912.

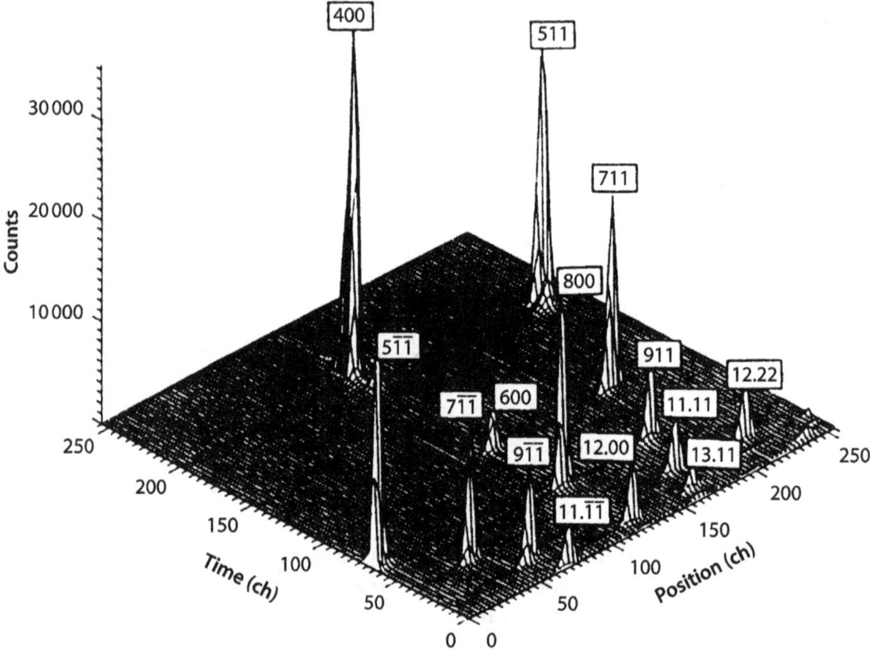

Fig. 3.86. Laue diffraction pattern of a Ge single crystal measured with TOF technique at the spallation source ISIS

3.12
Conclusion

Careful experiments with powder diffractometers give well resolved diffraction patterns. Since a diffraction diagram contains all the crystallographic information to obtain good results is solely a question of sensible analysis. This has to begin with *profile analysis* in order to determine a reliable *profile function*, followed by a *profile fitting* procedure. Many programs are available worldwide. *Profile fitting* provides the peak positions for lattice constants and symmetry determination, and integrated intensities for any crystallographic calculation, if the Two Step method is applied. This method separates the problem into two from each other completely independent steps and opens the way especially well. While crystal structures can be refined by several ways, Fourier calculations can only be done if the integrated peaks are available independently. At this stage powder diffraction is competitive with single crystal work. In recent years this method has been extended even to crystal structure analysis by direct methods with great success.

References

Ahtee M, Unonius L, Nurmela M, Suortti P (1984) A Voigtian as profile shape function in Rietveld refinement. J Appl Cryst 17:352–357

Ahtee M, Nurmela M, Suortti P, Järvinen M (1989) Correction for preferred orientation in Rietveld refinement. J Appl Cryst 22:261–268

Akima II (1970) J Assc For Comp Math 17,:589

Albinati A, Willis BTM (1982) The Rietveld method in neutron and X-ray powder diffraction. J Appl Cryst 15:361–374

Alexander LE, Klug HP (1948) Basic aspects of X-ray absorption in quantitative diffraction analysis of powder mixtures. Anal Chem 20:886–889

Altomare A, Carrozzini B, Giacovazzo C, Guagliardi A, Moliterni AGG, Rizzi R (1996a) Solving crystal structures from powder data. I. The role of prior information in the Two-Stage method. J Appl Cryst 29:667–673

Altomare A, Foadi J, Giacovazzo C, Guagliardi A, Moliterni AGG (1996b) Solving crystal structures from powder data. II. Pseudotranslational symmetry and powder-pattern decomposition. J Appl Cryst 29:674–681

Appleman DE, Evans HT Jr (1973) US Geological Survey Computer Contribution 20. US National Technical Information Service Document PB2-16188

Attfield JP, Sleight AW, Cheetham AK (1986) Structure determination of α-CrPO$_4$ from powder synchrotron X-ray data. Nature 322:620–622

Bacon GE (1975) Neutron diffraction. Oxford University Press, Oxford, p 152

Barlow W (1883) Probable nature of the internal symmetry of crystals. Nature 29:186–188, 205–207

Barlow W (1897) A mechanical cause of homogeneity of structure and symmetry. Proceedings of the Royal Dublin Society

Berg JE, Werner PE (1977) On the use of Guinier-Hägg film data for structure analysis. Z Kristallogr 145:310–320

Bernard P, Louer M, Louer DJ (1991a) Crystal structure determination of Zr(OH)$_2$(NO$_3$)$_2 \cdot$ 4.7 H$_2$O from X-Ray powder diffraction data. Solid State Chemistry 94:27–35

Bernard P, Louer M, Louer D (1991b) Solving the crystal structure of Cd$_6$(OH)$_9$(NO$_3$)$_2 \cdot$ 3 H$_2$O from powder diffraction data. A comparison with single crystal data. Powder Diffr 6:10–15

Boultif A, Louer D (1991) Indexing of powder diffraction patterns for low-symmetry lattices by the successive dichotomy method. J Appl Cryst 24:987–993

Bragg WL (1913) The structure of some crystals as indicated by their diffraction of X-rays. P Roy Soc Lond A 89:248–277

Bragg WH, Bragg WL (1913) The structure of the diamond. P Roy Soc Lond A 89.277–291

Bragg WH, Bragg WL (1933) The crystalline state. G. Bell and Sons, London

Brindley GW (1945) Philos Mag 36:347–369

Burla C, Camalli M, Cascarano G, Giacovazzo C, Polidori G, Spagna R, Viterbo D (1989) SIR88 – a direct-methods program for automatic solution of crystal structures. J Appl Cryst 22:389–393

Cagliotti G, Paoletti A, Ricci FP (1958) Choice of collimators for a crystal spectrometer for neutron diffraction. Nucl Instrum Methods 3:223–228

Carrozzini B, Giacovazzo C, Guagliardi A, Rizzi R, Burla MC, Polidori G (1997) Solving crystal structures from powder data. III. The use of the probability distributions for estimating the $|F|$'s. J Appl Cryst 30:92–97

Cascarano G, Favia L, Giacovazzo C (1992) SIRPOW.91 – A direct methods package optimized for powder data. J Appl Cryst 25:310–317

Cernik RJ, Murray PK, Pattison P, Fitch AN (1990) A two-circle powder diffractometer for synchrotron radiation with a closed loop encoder feedback system. J Appl Cryst 23:292–296

Cernik RJ, Cheetham AK, Prout CK, Watkin DJ, Wilkinson AP, Willis BTM (1991) The structure of cimetidine ($C_{10}H_{16}N_6S$) solved from synchrotron-radiation X-ray powder diffraction data. J Appl Cryst 24:222–226

Cheary R, Coelho A (1992) A fundamental parameters approach to X-ray line-profile fitting. J Appl Cryst 25:109–121

Cheetham AK, Taylor JC (1977) Profile analysis of powder neutron difraction data: Its scope, limitations, and applications in solid state chemistry. J Solid State Chem 21:253–275

Clearfield A, McCusker LB, Rudolf PR (1984) Crystal structure from powder data. 1. Crystal structure of $ZrKH(PO_4)_2$. Inorg Chem 23:4679–4582

Coelho AA (2000) Whole-profile structure solution from powder diffraction data using simulated annealing. J Appl Cryst 33:899–908

Coelho AA (2003) Indexing of powder diffraction patterns by iterative use of singular value decomposition. J Appl Cryst 36:86–95

Cooper MJ (1983) The validity of the Rietveld method. Z Kristallogr 164:157–158

Cooper MJ, Sakata M, Rouse KD (1980) The determination of structural parameters and their standard deviations from powder diffraction patterns. Proceedings of Symposium on Accuracy in Powder Diffraction held at NBS, Gaithersburg, MD (NBS Special PubliCation 567:167–187)

Cooper MJ, Rouse KD, Sakata M (1981) An alternative to the Rietveld profile refinement method. Z Kristallogr 157:101–117

Coppens P (1971) Experimental electron densities and chemical bonding. Angew Chem Int Edit 208:32–40

Cox DE, Hastings JB, Thomlinson W, Prewitt CT (1983) Application of synchrotron radiation to high resolution powder diffraction and Rietveld refinement. Nucl Instrum Methods 208:573–578

Cromer DT, Liberman D (1970) Relativistic calculation of anomalous scattering factors for X-rays. J Chem Phys 53:1891–1898

Cromer DT, Liberman D (1981) Anomalous dispersion calculations near to and on the long-wavelength side of an absorption edge. Acta Cryst A 37:267–268

Debye P, Scherrer P (1916a) Kgl. Ges. d. Wiss., Göttingen, Dec 1915

Debye P, Scherrer P (1916b) Interferenzen an regellos orientierten Teilchen im Röntgenlicht. Phys Z 17:277–283

Dollase WA (1986) Correction of intensities for preferred orientation in powder diffractometry: Application of the March model. J Appl Cryst 19:267–272

Engler O, Palacios J, Schäfer W, Jansen E, Lücke K, Will G (1993) Comparison between texture results obtained by X-ray and neutron measurements. Texture Microstruct 21:195–206

Estermann MA, Lynne B, McCusker LB, Baerlocher C (1992) *Ab initio* structure determination from severely overlapping powder diffraction data. J Appl Cryst 25:539–543

Ewald PP (1962) Fifty years of X-ray diffraction. International Union of Crystallography, Utrecht, Netherlands, pp 31–73

Ferrari M, Lutterotti P, Mathies P, Polonioli P, Wenk HR (1996) Mat Sci Forum 228–231:83

Geller S (1967) Crystal chemistry of garnets. Z Kristallogr 125:1–47

Giacovazzo C (1980) Direct methods in crystallography. Academic Press, London

Giacovazzo C (ed) (1992) Fundamentals of crystallography. International Union of Crystallography, Oxford University Press

Giacovazzo C (1996) Direct methods and powder data: State of the art and perspectives. Acta Cryst A 52:331–339

Hanawalt JD, Rinn HW, Frevel LK (1938) Chemical analysis by X-ray diffraction – classification and use of X-ray diffraction patterns. Ind Eng Chem Anal Ed 10:457–512

Hamilton WC (1964a) Program No. 313 (POWLS). ACA Computer Program Listing, November 1961

Hamilton WC (1964b) Statistics in physical science. Ronald Press, New York

Hart M, Cernik RJ, Parrish W, Toraya H (1990) Lattice parameter determination for powders using synchrotron radiation. J Appl Cryst 23:286–291

Hastings JB, Thomlinson W, Cox DE (1984) Synchrotron X-ray powder diffraction. J Appl Cryst 17:85–95

Haywood BC, Shirley R (1977) The structure of tetraiodoethylene at 4 K. Acta Cryst B 33:1765–1773

Hauptmann H, Karle J (1956) Structure invariants and semiinvariants for non-centrosymmetric space groups. Acta Cryst 9:45–55

Hill JR (1984) X-ray powder diffraction profile refinement of synthetic hercynite. Am Mineral 69:937–942

Hill JR (1991) Expanded use of the Rietveld method in studies of phase abundance in multiphase mixtures. Powder Diffr 6:74–77

Hill JR, Fischer RX (1990) Profile agreement indices in Rietveld and pattern-fitting analysis. J Appl Cryst 23:462–468

Hill JR, Howard CJ (1987) Quantitative phase analysis from neutron powder diffraction data using the Rietveld method. J Appl Cryst 20:467–474

Hill JR, Madsen IC (1991) Rietveld analysis using para-focusing and Debye-Scherrer geometry data collected with a Bragg-Brentano diffractometer. Z Kristallogr 196:73–92

Höffner C, Will G (1991) PC-profile analysis of peak clusters in angle and energy dispersive powder diffractometry. Mater Sci Forum 79–82: 91–98

Höfler S, Schäfer W, Will G (1985) Texture measurements at the neutron diffractometer in Jülich. In: Bunge HJ (ed) Experimental techniques of texture analysis. DMG Informationsgesellschaft Verlag, pp 241–251

Höfler S, Will G, Hamm H-M (1988) Neutron diffraction pole figure measurements on iron meteorites. Earth Planet Sc Lett 90:1–10

Höfler S, Schäfer W, Will G (1989) Neutron diffraction pole figure measurements for the microstructure analysis of meteorites. Physica B 156/157:675–677

Honkimäki V, Suortti P (1992) Whole pattern fitting in energy-dispersive powder diffraction. J Appl Cryst 25:97–104

Howard CJ (1982) The approximation of asymmetric neutron powder diffraction peaks by sums of Gaussian. J Appl Cryst 15:615–620

Huang TC, Parrish W (1975) Accurate and rapid reduction of experimental X-ray data. Appl Phys Lett 27:123–124

Huang TC, Hart M, Parrish W, Masciocchi N (1987) Line-broadening analysis of synchrotron X-ray diffraction data. J Appl Phys 61:2813–2816

Hüpen H, Will G, Höffner C, Elf F (1991) X-ray diffraction of CuS_2 under high pressure. Mat Sci Forum 79–82:697–702

Hull AW (1917) Phys Rev 9:84

Jansen E, Schäfer W, Will G (1985) Application of profile methods in texture measurements using position-sensitive detectors. In: Bunge HJ (ed) Experimental techniques of texture analysis. DMG Informationsgesellschaft Verlag, pp 229–240

Jansen E, Schäfer W, Will G. (1988) Profile fitting and the Two-Stage method in neutron powder diffractometry for structure and texture analysis. J Appl Cryst 21:228–239

Jansen E, Schäfer W, Will G (1989) International Workshop on the Rietveld Method. Netherland Energy Research Foundation, ECN, Petten, p 64 (see also Jansen et al. 1994a).

Jansen E, Schäfer W, Will G (1994a) R-values of powder diffraction data using Rietveld refinement. J Appl Cryst 27:492–496

Jansen E, Schäfer W, Will G (1994b) PC environment for commonly used programs in powder structure refinement. Mat Sci Forum 166–169:187–192

Jansen J, Peschar R, Schenk H (1992) On the determination of accurate intensities from powder diffraction data. I. Whole-pattern fitting with a least-squares procedure. J Appl Cryst 25:231–236 Järvinen M, Merisalo M, Pesonen A, Inkinen O (1970) Correction of integrated X-ray intensities for preferred orientation in cubic powders. J Appl Cryst 3:313–318

Jorgensen JD, Rotella FJ (1982) High-resolution time-of-flight powder diffractometer at the ZING-P' pulsed neutron source. J Appl Cryst 15:27–34

Karle J, Hauptmann H (1950) The phases and magnitudes of the structure factors. Acta Cryst 3:181–187

Khattak CP, Cox DE (1977) Profile analysis of X-ray powder diffractometer data: Structural refinement of $La_{0.75}Sr_{0.25}CrO_3$. J Appl Cryst 10:405–411

Kirfel A, Will G (1980) Bonding in $S_2O_6^{2-}$: Refinement and pictorial representation from an X-ray diffraction study of $Na_2S_2O_6 \cdot 2\,H_2O_6$ and $Na_2S_2O_6 \cdot 2\,D_2O_6$. Acta Cryst B 36:512–523

Klug HP, Alexander LE (1974) X-ray diffraction procedures for polycrystalline and amorphous materials. J. Wiley, New York

Kockelmann W (1995) Bestimmung kommensurabler und inkommensurabler Magnetstrukturen aus Neutronenpulverdiffraktometrie an Verbindungen der Selten-Erd (R) Systeme RXO_4, RFe_5Al_7 und $R(Co,Ni)C_2$. JÜL Report, JÜL-3025, Februar 1995, ISSN 0944-2952

Kockelmann W, Schäfer W, Will G (1992a) Low temperature neutron diffraction investigation of the complex incommensurate magnetic ordering of $TbAsO_4$. J Phys Chem Solids 53:913–921

Kockelmann W, Schäfer W, Will G (1992b) The incommensurate magnetic structure of $TbAsO_4$ analyzed by profile fitting. Physica B 180/181:68–70

Kockelmann W, Schäfer W, Will G, Fischer P, Gal J (1994) Neutron diffraction study of the ferrimagnetic structures of RFe_5Al_7 compounds with R = Tb, Dy, Ho, Er, Tm. J Alloy Compd 207/208:311–315

Kockelmann W, Jansen E, Schäfer W, Will G (1995) IC-POWLS: A program for calculation and refinement of commensurate and incommensurate structures using powder diffraction. Report JÜL-3025, KFA Jülich, Germany

Kurdjumow G, Sachs G (1930) Über den Mechanismus der Stahlhärtung. Z Phys 64:325–343

Lager AG, Ross FK, Rotella FJ, Jorgensen JDT (1981) Neutron powder diffraction of forsterite, Mg_2SiO_4: A comparison with single-crystal investigations. J Appl Cryst 14:137–139

Larson AC, Von Dreele RB (1994) General structure analysis system (GSAS). Los Alamos National Laboratory Report LAUR 86.748

Lauterjung J, Will G, Hinze E (1985) A fully automatic peak-search program for the evaluation of Gauss-shaped diffraction patterns. Nucl Instrum Meth A 239:281–287

Le Bail A, Duroy H, Fourquet JL (1988) *Ab-initio* determination of $LiSbWO_4$ by X-ray powder diffraction. Mat Res Bull 23:447–452

Le Page Y, Donnay G (1976) Refinement of the crystal structure of low-quartz. Acta Cryst B 32:2456–2459

Lehmann MS, Christensen AN, Fjellvag H, Feidenhans LR, Nielsen M (1987) Structure determination by use of pattern decomposition and the Rietveld method on synchrotron X-ray and neutron powder data; the structure of $Al_2Y_4O_9$ and I_2O_4. J Appl Cryst 20:123–139

Levien L, Prewitt C, Weidner DJ (1980) Structure and elastic properties of quartz at pressure. Am Mineral 65:920–930

Levy JH, Taylor JC, Wilson PW (1975a) The structure of fluorides. IX. The orthorhombic form of molybdenum hexafluoride. Acta Cryst B 31:398–401

Levy JH, Taylor JC, Wilson PW (1975b) The structure of uranium(III) triiodide by neutron diffraction. Acta Cryst B 30:880–882

Loopstra BO, Boldrini P (1966) Neutron diffraction investigation of WO_3. Acta Cryst 21:158–162

Loopstra BO, Rietveld HM (1969) The structure of some alkaline-earth metal uranates. Acta Cryst B 25:787–791

Louer D, Louer MJ (1972) Methode d'essais et erreurs pour l'indexation automatique des diagrammes de poudre. J Appl Cryst 5:271–275

Louer D, Louer MJ (1987) Crystal structure of $Nd(OH)_2NO_3 \cdot H_2O$ completely solved and refined from X-ray powder diffraction. J Solid State Chem 68:292–299

Louer D, Vargas R (1982) Indexation automatique des diagrammes de poudre par dichotomies successives. J Appl Cryst 15:542–545

Louer D, Louer M, Touboul MJ (1992) Crystal structure determination of lithium diborate hydrate, $LiB_2O_3(OH) \cdot H_2O$ from X-ray powder diffraction data collected with a curved position-sensitive detector. Acta Cryst 25:617–623

Lutterotti L, Scardi P (1990) Simultaneous structure and size-strain refinement by the Rietveld method. J Appl Cryst 23:246–252

Lutterotti L, Matthies S, Wenk HR, Schulz AS, Richardson JW Jr (1997) Combined texture and structure analysis of deformed limestone from time-of-flight neutron diffraction spectra. J Appl Phys 81:594–600

Madsen IC, Hill RJ (1991) QPDA – A user-friendly, interactive program for quantitative phase and crystal size/strain analysis of powder diffraction data. Powder Diffr 5:105–199

Madsen IC, Scarlett NVY, Cranswick LMD, Lwin T (2001) Outcomes of the International Union of Crystallography Commission on Powder Diffraction Round Robin on quantitative phase analysis: samples 1a to 1h. J Appl Cryst 34:409–426

Main P, Lessinger L, Woolfson MM, Germain G, Declercq J-P (1977) MULTAN77. A system of computer programs for the automatic solution of crystal structures from X-ray diffraction data. University of York, England and Louvain, Belgium

Main P, Fiske SJ, Hull SE, Lessinger L, Germain G, Declerq JP, Wolfson MM (1980) MULTAN80: A system of computer programs for the automatic solution of crystal structures from X-ray diffraction data. University of York and Louvain

Malmros G, Thomas JO (1977) Least-squares structure refinement based on profile analysis of powder film intensity data measured on an automatic microdensitometer. J Appl Cryst 10:7–11

Malmros G, Werner PE (1973) Automatic densitometer measurements of powder diffraction photographs. Acta Chem Scand 27:493–502

March A (1932) Mathematische Theorie der Regelung nach der Korngestalt bei affiner Deformation. Z Kristallogr 81:285–297

McCusker LB (1988) The *ab initio* structure determination of sigma-2 (a new clathrasil phase) from synchrotron powder diffraction data. J Appl Cryst 21:205–310

McCusker LB, Von Dreele RB, Cox DE, Louer D, Scardi P (1999) Rietveld refinement guidelines. J Appl Cryst 32:36–50

Merz P, Jansen E, Schäfer W, Will G (1990) PROFAN-PC: A PC-program for powder peak profile analysis. J Appl Cryst 23:444–445

Nishiyama Z (1934) X-ray investigation of the mechanism of the transformation from face-centered cubic lattice to body-centered cubic. Sci Rep Tohuku Imp Univ Tokio 23:637–664

Nover G, Will G (1981) Structure refinement of seven natural olivine crystals and the influence of the oxygen partial pressure on the cation distribution. Z Kristallogr 190:277–285

Parrish W (1960) Results of the IUCr precision lattice parameter project. Acta Cryst 13:838–850

Parrish W (1965) X-ray analysis papers. Centrex Publishing Comp., Eindhoven

Parrish W, Huang TC (1980) Accuracy of the profile fitting method for X-ray polycrystalline diffracto-metry. Proceedings of the Symposium on Accuracy in Powder Diffraction. Nat. Bur. of Stand., Washington D.C. (Special Publication 567:95–119)

Parrish W, Huang TC (1983) Accuracy and precision of intensities in X-ray polycrystalline diffraction. Adv X Ray Anal 26:35–44

Parrish W, Huang TC, Ayers GL (1976) Profile fitting: A powerful method of computer X-ray instrumentation and analysis. Trans Am Cryst Assoc 12:55

Parrish W, Hart M, Huang TC, Bellotto M (1987) Lattice parameter determination using synchrotron powder data. Adv X Ray Anal 30:373–382

Paterson MS (1950) Proc Phys Soc London Sect A 63:477–482

Pawley GS (1980) EDINP, the Edinburgh powder refinement program. J Appl Cryst 13:630–633

Pawley GS (1981) Unit cell refinement from powder diffraction scans. J Appl Cryst 14:357–361

Peplinski B, Kleberg R, Bergmann J, Wenzel J (2004) Quantitative phase analysis using the Rietveld method – Estimates of possible problems based on two interlaboratory comparisons. Mater Sci Forum 443–444:45–50

Phillips JC, Hodgson KO (1980) The uses of anomalous scattering effects to phase diffraction patterns from macromolecules. Acta Cryst A 36:856–864

Plevert J, Louer M, Louer D (1989) The ab $initio$ structure determination of $Cd_3(OH)_5(NO_3)$ from X-ray powder diffraction data. J Appl Cryst 22:470–475

Press W, Flannery B, Teukolsky S, Vetterling W (1988) Numerical revues in C. Cambridge Library

Prince E (1985) Structure in statistics in crystallography. Academic Press, pp 95–103

Rietveld HM (1966) The crystal structures of some alkaline earth metal uranates of the type M_3UO_6. Acta Cryst 20:508–513

Rietveld HM (1967) Line profiles of neutron powder-diffraction peaks for structure refinement. Acta Cryst 22:151–152

Rietveld HM (1969) A profile refinement method for nuclear and magnetic structures. J Appl Cryst 2:65–71

Rius J, Plana F, Palanques A (1987) Standardless X-ray diffraction method for quantitative analysis of multiphase mixtures. J Appl Cryst 20:457–460

Rodriguez-Carvajal J (2001) Recent developments of the program FULLPROF. International Union of Crystallography, Newsletter No. 26, 26 December 2001

Sakata M, Cooper MJ (1979) An analysis of the Rietveld profile refinement method. J Appl Cryst 12:554–536

Savitzky A, Golay MJE (1964) Smoothing and differentiation of data by simplified least squares procedures. Anal Chem 36:1627–1639

Scarlett NVY, Madsen IC, Cranswick LMD, Lewin T, Groleau E, Stephenson G, Aylmore M, Agron-Olshina N (2002) Outcomes of the International Union of Crystallography Commission on Powder Diffraction Round Robin on quantitative phase analysis: samples 2, 3, 4, synthetic bauxite, natural granodiorite and pharmaceuticals, J Appl Cryst 35:383–400

Schäfer W, Will G, Müller-Vogt G (1979) Refinement of the crystal structure of terbium arsenate $TbAsO_4$ at 77 K and 5 K by profile analysis from neutron diffraction powder data. Acta Cryst B 35:588–592

Schäfer W, Höfler S, Will G (1988) Applications of neutron diffraction pole figure measurements on polycrystalline and monocrystalline metallic samples. Examples from the four-circle neutron diffractometer in Jülich. Texture Microstruct 8/9:457–466

Schäfer W, Merz P, Jansen E, Will G (1991) Neutron diffraction texture analysis of multiphase and low-symmetry materials using the position-sensitive detector JULIOS and peak deconvolution methods. Texture Microstruct 14–18:65–71

Schäfer W, Jansen E, Merz P, Will G, Wenk HR (1992a) Neutron diffraction texture investigation on deformed quartzites. Physica B 180/181:1035–1038

Schäfer W, Jansen E, Skowronek R, Will G, Knight KS, Finney JL (1992b) Time-of-flight powder diffraction test experiments at a pulsed neutron beam using the position-sensitive detector JULIOS. Nucl Instrum Meth A 317:202–212

Schäfer W, Jansen E, Will G (1993) Angle dispersive time-of-flight diffraction in a pulsed beam: An efficient technology to exploit the thermal-neutron spectrum – design of a JULIOS diffractometer and experimental tests. J Appl Cryst 26:660–669

Schäfer W, Jansen E, Will G, Kotsanidis PA, Yakinthos JK, Tietze-Jaensch H (1994) Crystalographic and magnetic structure of TbPtGa. J Alloy Compd 206:225–229

Schäfer W, Jansen E, Skowronek R, Will G, Kockelmann W, Schmidt W, Tietze-Jaensch H (1995) Setup and use of the Rotax instrument at ISIS as an engle-dispersive neutron powder and texture diffractometer. Nucl Instrum Methods A364:179–185

Scott HG (1983) The estimation of standard deviations in powder diffraction Rietveld refinements. J Appl Cryst 16:159–163

Sheldrick GM (1976) Program for crystal structure determination. University of Cambridge, England

Shull CG, Smart JS (1949) Detection of antiferromagnetism by neutron diffraction. Phys Rev 76:1256–1257

Shull CG, Strausser WA, Wollan EO (1951) Neutron diffraction by paramagnetic and antiferro-magnetic substances. Phys Rev 83:333–345

Smith GS, Alexander LE (1963) Refinement of the atomic parameters of α-quartz. Acta Cryst 16:462–471

Smith DK, Gorter SJ (1991) Powder diffraction program information. 1990 program list. Appl Cryst 24:369–402

Sonnefeld EJ, Visser JW (1975) Automatic collection of powder data from photographs. J Appl Cryst 8:1–7

Steenstrup S (1981) A simple procedure for fitting a background to a certain class of measured spectra. J Appl Cryst 14:226–229

Stewart JM, Hall SR (1988) XTAL user's manual. Computer Science Center, University of Maryland (Technical Report TR1364.2); also in: Crystallographic computing. Oxford University Press, pp 325–353

Stout GH, Jensen LH (1970) X-ray structure determination. The Macmillan Company, London

Suortti P, Jennings LD (1977) Accuracy of structure factors from X-ray powder intensity measurements. Acta Cryst A33:1012–1027 (Commission on Crystallographic Apparatus, International Union of Crystallography)

Suortti P, Ahtee M, Unonius L (1979) Voigt function fit of X-ray and neutron powder diffraction profiles. J Appl Cryst 12:365–369

Taylor JC (1987) A comparison of profile decomposition and Rietveld methods for structural refinement with powder diffraction data. Z Kristallogr 181:151–160

Taylor JC (1991) Computer programs for standardless quantitative analysis of minerals using the full powder diffraction profile. Powder Diffr 6:2–9

Taylor JC, Matulis CE (1991) Absorption contrast effects in the quantitative XRD phase analysis of powders by full multiphase profile refinement. J Appl Cryst 24:14–17

Taylor JC, Pescover SR (1988) Quantitative analysis of phases in zeolite bearing rocks from full X-ray diffraction profiles. Austr J Phys 41:323–335

Taylor JC, Wilson PW (1974a) The structure of β-tungsten hexachloride by powder neutron and X-ray diffraction. Acta Cryst B 30:1216–1220

Taylor JC, Wilson PW (1974b) Neutron and X-ray powder diffraction studies of the structure of uranium hexachloride. Acta Cryst B 30:1481–1484

Taylor JC, Wilson PW (1974c) The structure of uranium tetrabromide by X-ray and neutron diffraction. Acta Cryst B 30:2664–2667

Taylor JC, Miller SA, Bibby DM (1986) A study of decomposition methods for refinement of H^+-ZSM5 zeolite with powder diffraction data. Z Kristallogr 176:183–192

Taupin D (1973) Automatic peak determination in X-ray powder patterns. J Appl Cryst 6:266–273

Templeton DH, Templeton LK, Phillips JC, Hodgson KO (1980) Anomalous scattering of X-rays by cesium and cobalt measured with synchrotron radiation. Acta Cryst A 36:436–442

Templeton LK, Templeton DH, Phizackerley RP, Hodgson KO (1982) L3-edge anomalous scattering by gadolinium and samarium measured at high resolution with synchrotron radiation. Acta Cryst A 38:74–78

Toraya H (1986) Whole powder pattern fitting without reference to a structural model: Application to X-ray powder diffraction data. J Appl Cryst 19:440–447

Toraya H (1988) The deconvolution of overlapping reflections by the procedure of direct fitting. J Appl Cryst 21:192–196

Toraya H (1989) Whole-pattern decomposition method. The Rigaku Journal 6:28–34

Toraya H, Yoshimura M, Somiya SA (1983) A computer program for the deconvolution of X-ray diffraction profiles with the composite of Pearson VII functions. J Appl Cryst 16:653–657

Valvoda V, Chladek M, Cerney R (1996) Joint texture refinement. J Appl Cryst 29:48–52

Vand V, Niggli A (1961) The use of a Monte Carlo method in X-ray structure analysis. In: Computing methods and the phase problem in X-ray crystal analysis. Pergamon Press, Oxford, pp 266–272

Visser JW (1969) A fully automatic program for finding the unit cell from powder data. J Appl Cryst 2:89–95

Von Dreele RB (1995) Abstract. European Powder Diffraction Conference EPDIC IV, Chester, England

Von Dreele RB (1997) Quantitative texture analysis by Rietveld refinement. J Appl Cryst 30:517–525

Von Dreele RB, Cheetham AK (1974) Proc Roy Soc A 338:311–326

Von Dreele RB, Jorgensen JD, Windsor CG (1982) Rietveld refinement with spallation neutron powder diffraction data. J Appl Cryst 15:581–589

Warren BE, Averbach BL (1950) The effect of cold-work distortion on X-ray patterns. J Appl Phys 21:595–599

Warren BE, Averbach BL (1952) The separation of stacking fault broadening in cold-work metals. J Appl Phys 23:1059

Wassermann G (1935) Über den Mechanismus der α-γ-Umwandlung des Eisens. Mitt K Wilh-Inst f Eisenforschung 17:149–155

Weiss Z, Krajicek J, Smrcok L, Fiala J (1983) A computer X-ray quantitative phase analysis. J Appl Cryst 16:493–497

Wenk HR, Grigull S (2003) Synchrotron texture analysis with area detectors. J Appl Cryst 36:1040–1049

Wenk HR, Kern H, Schäfer W, Will G (1984) Comparison of neutron and X-ray diffraction in texture analysis of deformed carbonate rocks. J Struct Geol 6:687–692

Wenk HR, Matthies S, Hemley RJ, Mao HK, Shu J (2000) The plastic deformation of iron at pressures of the Earth's inner core. Nature 405:1044–1047

Werner P-E (1964) Trial-and-error computer methods for the indexing of unknown powder patterns. Z Kristallogr 120:375–387

Werner P-E (1976) On the determination of unit cell dimensions from inaccurate powder diffraction data. J Appl Cryst 9:216–219

Werner P-E, Salome S, Malmros G, Thomas JO (1979) Quantitative analysis of multicomponent powders by full-profile refinement of Guinier-Hägg X-ray film data. J Appl Cryst 12:107–109

Werner P-E, Erikson, Westdahl (1985) TREOR, a semi-exhausitve trial-and error powder indexing program for all symmetries. J Appl Cryst 18: 367–370

Wiles DB (1982a) Program for Rietveld analysis of X-ray and neutron powder diffraction patterns: DBW3.2. Georgia Institut of Technology, Atlanta, USA

Wiles H (1982b) The deconvolution of overlapping reflections by the procedure of direct fitting. J Appl Cryst 15:430–438

Wiles DB, Young RA (1981) A new computer program for Rietveld analysis of X-ray powder diffraction patterns. J Appl Cryst 14:149–151

Will G (1968) Lokale Verteilung der magnetischen Momentdichte in Kristallen aus Neutronenbeugungsuntersuchungen. Z Angew Phys 24:260–269

Will G (1969a) Kristallstrukturanalyse und Neutronenbeugung. I: Neutronenbeugung an Atomkernen. Angew Chem 81:984–995

Will G (1969b) Kristallstrukturanalyse und Neutronenbeugung. II: Neutronenbeugung und magnetische Strukturen. Angew Chem 81:984–995

Will G (1969c) Magnetsche Strukturuntersuchungen und Formfaktormessungen an Seltenen Erden mit Neutronenbeugung. Z Angew Phys 26:67–76

Will G (1971) Neutronenbeugung und Spinstrukturen Seltener Erden und intermetallischer 3d-4f-Verbindungen. Z Angew Phys 32:1–8

Will G (1972) Thermal neutron scattering by magnetic interaction. Ann Phys 7:371–405

Will G (1979) POWLS: A powder least-squares program. J Appl Cryst 12:483–485

Will G (1988) Crystal structure analysis and refinement using integrated intensities from accurate profile fits. Austr J Phys 41:283–296

Will G (1989) Crystal structure analysis from powder diffraction data. Z Kristallogr 188:169–186

Will G, Berndt H (2001) Kinetics of the pressure induced first-order phase transformation of RbJ. High Pressure Res 21:215–225

Will G, Höffner C (1992) HFIT, a Program for interactive analysis of powder diffraction data. Adv X Ray Anal 35:571–572

Will G, Lauterjung J (1987) The kinetics of the pressure induced olivine-spinel phase transition. In: Manghnani MH, Syono Y (eds) High pressure research in mineral physics. American Geophysical Union Washington, D.C., pp 177–186

Will G, Frazer BC, Cox DE (1965) The crystal structure of MnSO$_4$. Acta Cryst 19:854–857

Will G, Nuding W, Hinze E (1980a) X-ray energy dispersive diffractometry using profile analysis for high pressure experiments. In: High pressure technology. Pergamon Press, New York (Proceedings of the 7th International AIRAPT Conference, Le Creusot, France, vol 1, pp 201–203)

Will G, Jansen E, Schäfer W (1980b) POWLS60: A program for refinement of powder diffraction data. Report JÜL-1646, KFA Jülich, Germany

Will G, Parrish W, Huang TC (1983a) Crystal structure refinement by profile fitting and least squares analysis of powder diffractomter data. J Appl Cryst 16:611–622

Will G, Jansen E, Schäfer W (1983b) POWLS80: A program for calculation and refinement of powder diffraction data. Report JÜL-1867, KFA Jülich, Germany

Will G, Lauterjung,J, Schmitz H, Hinze E (1984) The bulk moduli of 3d-transition element pyrites measured with synchrotron radiation in a new belt type apparatus. Mat Res Soc Symp Proc 22:49–52

Will G, Höfler S, Schäfer W (1986a) Application of neutron diffraction in texture analysis. Physica 136B:473–476

Will G, Höfler S, Schäfer W (1986b) Texture analysis by neutron diffraction. Mat Sci Forum 27/28:457–464

Will G, Masciocchi N, Hart M, Parrish W (1987a) Ytterbium LIII-edge anomalous scattering measured with synchrotron radiation powder diffraction. Acta Cryst A 43:677–683

Will G, Masciocchi N, Parrish W, Hart M (1987b) Refinement of simple crystal structures from synchrotron radiation powder diffraction data. J Appl Cryst 20:394–401

Will G, Bellotto M, Parrish W, Hart M (1988a) Crystal structure of quartz and magnesium germanate by profile analysis of synchrotron radiation high resolution powder data. J Appl Cryst 21:182–191

Will G, Höfler S, Schäfer W (1988b) Texture analysis by neutron diffraction. Mat Sci Forum 27/28:457–464

Will G, Schäfer W, Merz P (1989) Texture analysis by neutron diffraction using a linear position sensitive detector. Texture Microstruct 10:375–387

Will G, Jansen E, Schäfer W (1990a) Structure refinements in chemistry and physics. A comparative study using the Rietveld and the Two-Step method. Adv X Ray Anal 33:261–268

Will G, Maschiocchi N, Parrish W, Lutz HD (1990b) Crystal structure analysis and refinement of CrMnInS$_4$ from powder diffraction data using synchrotron radiation. Z Kristallogr 190:277–285

Will G, Merz P, Schäfer W, Dahms M (1990c) Application of position sensitive detectors for neutron diffraction texture analysis of hematite ore. Adv X Ray Anal 33:277–283

Will G, Jansen E, Schäfer W (1992a) Texture analysis of bulk samples by neutron diffraction using a position sensitive detector. Adv X Ray Anal 35:285–291

Will G, Huang TC, Sequeda F (1992b) Determination of crystal structure and cation distribution in thin films. Adv in X Ray Anal 35:151–157

Will G, Schäfer W Jansen E (1994) JULIOS: A position-sensitive neutron detector and its application to white beam time-of-flight diffraction. Adv X Ray Anal 37:375–384

Wilson AJC (1963) Mathematical theory of X-ray powder diffractometry. Philips Technical Library, Eindhoven

Wilson AJC (1976) Statistical bias in least-squares refinement. Acta Cryst A 32:994–996

Wilson AJC (1980) Accuracy in methods of lattice-parameter measurements. Nat Bur Stand Spec Pub 567:325

Worlton TG, Jorgensen JD, Beyerlein RA, Decker DL (1976) Multicomponent profile refinement of time-of-flight neutron diffraction data. Nuclear Instrum Methods 137:331–337

Young RA, Wiles DB (1982) Profile shape functions in Rietveld refinements. J Appl Cryst 15:430–438

Young RA, Mackie PE, Von Dreele RB (1977) Application of the pattern fitting structure-refinement method of X-ray powder diffractometer patterns. J Appl Cryst 10:262–269

Young RA, Prince E, Sparks RA (1982) Suggested guidlines for the publication of Rietveld analyses and pattern decomposition studies. J Appl Cryst 15:357–359

Zachariasen WH (1949) Crystal chemical studies of the 5f-series ekements. XI. The crystal structure of α-UF$_5$ and β-UF$_5$. Acta Cryst 2:296–298

Appendix

Appendix of an Example for Quartz

The data set consits of 35 observations coming from 54 Miller indices. The observations are based on the publication by G. Will, W. Parrish and T.C. Huang "*Crystal structure refinement by profile fitting and least squares analysis of powder diffractometer data*" (J Appl Cryst 16:611–622) (Will et al. 1983a).

The refinement is performed in 4 cycles dropping the Bragg-R-value from 31.5 to 3.5% in the second cycle and to $R = 0.97\%$ in the third cycle. The fourth cycle gave no further improvement.

IV.   INPUT DESCRIPTION

Card 1   FORMAT(18A4) Title

   Col. 1-72: alphanumeric text

Card 2   FORMAT(24I3) Control switches

   Col. 1-3 : INTEG(1) > 0   number of atoms in asymmetric unit
   Col. 2-6 : INTEG(2) > 0   number of form factor tables
                             being read from tape (see Card 7)
                     = 0   no form factor tables
                             (for neutron diffraction)
                     < 0   number of form factor tables
                             to be supplied on cards (see Card 7)
   Col. 7-9 : INTEG(3) = 0   LP-correction for X-rays (normal case
                             for graphite monochr. angle = 26.55 deg.)
                     = 1   LP-correction for X-rays (general)
                             for two monochromators. (see Card 6)
                     = 2   L-correction for neutrons
                             cylindrical sample
                     = 3   L-correction for neutrons
                             flat sample
                             (Definition of LP- and L-corrections
                             see CHAPTER III)
   Col.10-12: INTEG(4) = 0   isotropic thermal motion (see Card 5)
                   $\neq$ 0   anisotropic thermal motion (see Card 5a)
   Col.13-15: INTEG(5)   Miller index h ⎤ of preferred orientation
   Col.16-18: INTEG(6)   Miller index k ⎬ direction. If all indices
   Col.19-21: INTEG(7)   Miller index l ⎦ $\geq$ 0, than pref. or.
                                    maximum is at planes of
                                    indexed zone. Else maxi-
                                    mum is at indexed plane.
   Col.22-24: INTEG(8) < 0   unit weights
                     = 0   diagonal weighting scheme
                     > 0   non-diagonal weighting scheme
   Col.25-27: INTEG(9)   background value for calculation of
                               standard deviations for observations
                             being supplied without sigma(obs).
                             Sigma = SQRT{I(obs) + INTEG(9)}
   Col.28-30: INTEG(10)= 0   no matrix output
                   $\neq$ 0   matrix output (due to parameter correl-
                             ations)
   Col.31-33: INTEG(11) ⎤
   Col.34-36: INTEG(12) ⎥
     "   "   ⎬   INTEG(11)-INTEG(24)
                           free for user's application
   Col.70-72: INTEG(24) ⎦

Card 3 Cell

1st version   FORMAT(A2,6X,8F8)

   Col. 1-2 : Radiation symbol: Mo,Cu,Ni,Co,Fe,Cr,Ag (default=Cu)
   Col. 9-16: a ⎤   Lattice parameters
   Col.17-24: b ⎬   either in Angstrom   (>1.0)
   Col.25-32: c ⎦   or in reciprocal Angstrom   (<1.0)
   Col.33-40: alpha ⎤   Angles
   Col.41-48: beta ⎬   either in degree
   Col.49-56: gamma ⎦   or as cosines
   Col.57-64: Minimum value (either d or 2theta)
   Col.65-72: Maximum value (either 2theta or d)

<u>2nd version</u>   FORMAT(9F8)

   Col. 1-8 : wavelength in Angstrom
   Col. 9-16: ⎫
     "   ⎬ as above
   Col.65-72: ⎭

Card 4   FORMAT(I4,14I3) Symmetry

   Col. 1-4 : space group number due to Int. Tables
               (second setting preceeded by a minus sign)
               Input of Laue group and nonextinction (Col. 5-56)
               is only required when deviations from space group
               symmetry are desired.
   Col. 5-7 : Laue group indication (number code in Appendix A)
   Col. 8-10: ⎫
   Col.11-13: ⎪ Nonextinction indication
     "   ⎬   (number code in Appendix A)
   Col.54-56: ⎭

Card(s) 5   FORMAT(2A4,6F8) Atomic parameter

   Col. 1-8 : Atom identification (alphanumeric) (Name)
   Col. 9-16: Form factor table number for X-rays (see Card 7),
              or scattering length b [10^{-12} cm] for neutrons
   Col.17-24: Site occupancy (number of symmetry positions)
   Col.25-32: x-position ⎫
   Col.33-40: y-position ⎬ fractional coordinates
   Col.41-48: z-position ⎭
   Col.49-56: Isotropic temperature factor B

For input of anisotropic temperature factors a
second parameter card 5a is necessary for each atom:

<Card(s) 5a> FORMAT(6F8) [only if INTEG(4) ≠ 0 on Card 2]

   Col. 1- 8: beta(1,1) ⎫
   Col. 9-16: beta(2,2) ⎪
   Col.17-24: beta(3,3) ⎪ Definition of beta(i,j) due to
   Col.25-32: beta(1,2) ⎬ exp{-h(i) h(j) beta(i,j)}
   Col.33-40: beta(1,3) ⎪
   Col.41-48: beta(2,3) ⎭

Card(s) 6   FORMAT(3X,4A4,24X,2F12) Overall parameter

   Col. 1-16: Parameter identification (alphanumeric)
   Col.17-24: Initial value
   Col.25-32: Initial increment for calculation of partial
              derivatives

The set of overall parameter cards is ended by a blanc card.
This blanc card ends the input of model parameters, if no
form factors are required.
In the actual version of subroutine FUNNY in POWLS-80 the
following sequence of overall parameters is assumed:
1. scale factor
2. overall temperature factor
3. preferred orientation parameter
4. monochromator angle of incident beam
5. monochromator angle of reflected beam

Monochromator angles are considered only for INTEG(3)=1
on Card 2

<u>**Card(s) 7**</u> <u>to be supplied only for INTEG(2)≠0 on Card 2</u>

<u>1st version</u>   <u>FORMAT(free)   Form factors [for INTEG(2).gt.0 on Card 2]</u>

Values   Notation(s) of form factor table array NNF(I),
separated   I=1,N with N=the number of tables needed (see
by blancs, Col.9–14) on Card(s) 5. The number code on
ending   tape (file FXORFL) is listed in Appendix B.
with slash

<u>2nd version</u>   <u>FORMAT(free) [for INTEG(2).lt.0 on Card 2]</u>

(As above) Sequence(s) of form factor values $S(i),i=1,32$
            in steps of 0.05 in $\sin\theta/\lambda$ starting with $S(1)$
            for $\sin\theta/\lambda=0.0$. Each sequence starts with a
            new card and is ended by a slash (/).

Now the structure model is completely described. With this
input the program may be started for the calculation of the
structure model. The output will provide a numbered (hkl)-list
with calculated intensities.

<u>**Card(s) 8**</u>   FORMAT(2F9,10I3,2X,6I3)   Observations

Col. 1– 8: Observed intensity
Col. 9–16: Sigma of observation
Col.17–19: ⎫ Number(s) of the (hkl)-reflection list due
 " 20–22: ⎪ to the structure model indicating those
 "   :   ⎬ reflections, which contribute to the
Col.46–48: ⎭ observation.
Col 49–51: ⎫ Percentage correlations to subsequent observa-
 " 52–55: ⎬ tions may be presented here, if available. It
 "   :   ⎭ is not required in order to refine a structure.

When starting the program at this point [INTEG(8)=0 on
Card 2], a comparative list of model intensities and
and observations is printed out. No refinement is
performed.

<u>**Card(s) 9**</u>   FORMAT(72I1)   Parameter refinement switches

Col. 1   : ⎫ 0 or 1 array ISEL(I), I=1,N , with N = the
 "   2   : ⎪ maximum number of parameters. The index I
     3   : ⎬ corresponds to the parameter sequence on the
            ⎪ input.
            ⎪ ISEL(I) = 0 : The parameter is not refined.
Col.72   : ⎭ ISEL(I) = 1 : The parameter is refined

One card has to be supplied for each refinement cycle.

V.   EXAMPLE (INPUT/OUTPUT) for Quartz, SiO_2, X-ray diffraction

Card Input

```
        ----+----1----+----2----+----3----+----4----+----5----+----6----+----7--
Card 1  QUARZ COLLECTED WITH A GRAPHIT MONOCHROMATOR (PUBLISHED IN /5/)
Card 2     2   2   0   1   0   0   1   1
Card 3  CU        4.91329 4.91329 5.40483 90.      90.      120.              100.
Card 4   154
Card 5  SI       2.       3.       .4697   .0       .1666667
     a  .0093    .0078    .0049    .0039  -.00001  -.00002
     5  O        1.       6.       .4135   .2669    .2857667
     a  .0190    .0144    .0083    .0106  -.0025   -.0035
Card 6  SCALE             1.       .001
     6  OTEMP             .0       .001
     6  PREFOR            .0       .001
(Blanc Card)
Card 7  21 34/
Card 8      2932      62  1
     8     14110     122  2  3
     8      1034      45  4
     8       927      43  5  6
     8       465      38  7
     8       699      41  8
     8       486      38  9 10
     8      1830      70 11                                  -48
     8        35      60 12
     8       556      39 13 14                               -24
     8       212      34 15 16
     8        31      32 17
     8      1313      48 18 19
     8       226      35 20
     8        59      11 21
     8       823      45 22 23                               -19  2
     8       946      62 24 25                               -36
     8       568      50 26 27
     8       283      35 28 29
     8       409      37 30 31
     8       223      34 32
     8       428      53 33 34                               -43
     8        95      51 35
     8       335      53 36                                  -41
     8       389      55 37
     8       273      35 38 39
     8        37      32 40 41
     0         3      31 42
     8        36      32 43 44
     8       451      38 45 46
     8        83      32 47
     8       205      34 48 49
     8       151      33 50 51
     8       234      35 52 53
     0       210      34 54
(Blanc Card)
Card 9    1  111 1   1111111111 1
     9    1  111 1   1111111111 1
     9    1  111 1   1111111111 1
     9    1  111 1   1111111111 1
```

V.   EXAMPLE (INPUT/OUTPUT) for Quartz, SiO_2, X-ray diffraction

Selected parts of original output

```
QUARZ COLLECTED WITH A GRAPHIT MONOCHROMATOR (PUBLISHED IN /5/)

NO. OF ATOMS IN ASSYM. UNIT.   2      PREF. ORIENTATION (HKL)... 0 0 1
NO. OF FORM FACTOR TABLES ..   2      VERSION OF WEIGHTING SCHEME.   1
VERSION OF LP-CORRECTION ...   0      BACKGROUND.................   0
ANISOTROPIC TEMP. FACTORS... YES      MATRIX OUTPUT..............  NO

CU  RADIATION: ALPHA1 = 1.54060 ALPHA2 = 1.54444
2 THETA RANGE FROM  0.0  TO 100.00 DEGREE

CELL                            RECIPROCAL CELL

A    4.9133                     A*    0.2350
B    4.9133                     B*    0.2350
C    5.4048                     C*    0.1850
ALPHA  90.00                    ALPHA*  90.00
BETA   90.00                    BETA*   90.00
GAMMA 120.00                    GAMMA*  60.00

SPACE GROUP  154          NUMBER OF GENERAL POSITIONS   6

PATCH WILL CHANGE PARAMETERS

ATOM    SF      OC      X       Y       Z       B11     B22     B33     B12     B13     B23

SI   2.00000 3.00000 0.46970 0.0     0.16667 0.00930 0.00780 0.00490 0.00390 -0.00001 -0.00002
O    1.00000 6.00000 0.41350 0.26690 0.28577 0.01900 0.01440 0.00830 0.01060 -0.00250 -0.00350

SCALE    1.0000               OTEMP    0.0
PREFOR   0.0

FORM FACTOR TABLES FROM TAPE

TABLE NO.  1          O      SCF

     8.000   7.796   7.250   6.482   5.634   4.814   4.094   3.492
     3.010   2.674   2.338   2.141   1.944   1.829   1.714   1.640
     1.566   1.514   1.462   1.418   1.374   1.335   1.296   1.258
     1.220   1.182   1.144   1.106   1.068   1.030   0.992   0.0

TABLE NO.  2          SI     SCF

    14.000  13.450  12.160  10.790   9.670   8.850   8.220   7.700
     7.200   6.720   6.240   5.775   5.310   4.890   4.470   4.110
     3.750   3.455   3.160   2.925   2.690   2.520   2.350   2.210
     2.070   1.970   1.870   1.790   1.710   1.655   1.600   0.0
```

MILLER PLANES AND CALCULATED INTENSITIES FROM STARTING PARAMETERS

PLANE NO	2 THETA	D	H	K	L	MULT	LP	!F!	I(CALC)
1	20.88	4.25503	1	0	0	6.	58.3882	16.0	89192.5
2	26.66	3.34329	1	0	1	6.	35.1302	25.5	136554.9
3	26.66	3.34329	0	1	1	6.	35.1302	38.7	316141.7
4	36.58	2.45665	1	1	0	6.	17.9496	17.3	32332.2
5	39.50	2.28122	1	0	2	6.	15.1971	17.5	27809.3
6	39.50	2.28122	0	1	2	6.	15.1971	7.2	4748.5
7	40.33	2.23646	1	1	1	12.	14.5299	9.2	14916.2
8	42.49	2.12752	2	0	0	6.	12.9659	17.5	23795.3
9	45.84	1.97967	2	0	1	6.	10.9819	12.2	9740.2
10	45.84	1.97967	0	2	1	6.	10.9819	8.2	4435.7
11	50.19	1.81780	1	1	2	12.	8.9980	23.4	58970.5
12	50.67	1.80161	0	0	3	2.	8.8107	9.1	1448.1
13	54.93	1.67165	2	0	2	6.	7.3835	8.7	3329.1
14	54.93	1.67165	0	2	2	6.	7.3835	18.0	14347.7
15	55.38	1.65903	1	0	3	6.	7.2524	1.5	96.2
16	55.38	1.65903	0	1	3	6.	7.2524	13.5	7884.7
17	57.29	1.60825	2	1	0	12.	6.7384	3.7	1079.7
18	60.02	1.54146	2	1	1	12.	6.0968	16.5	19060.4
19	60.02	1.54146	1	2	1	12.	6.0968	17.8	23263.2
20	64.10	1.45281	1	1	3	12.	5.3094	11.3	8130.7
21	65.85	1.41834	3	0	0	6.	5.0244	8.2	2045.0
22	67.81	1.38203	2	1	2	12.	4.7378	10.6	6394.0
23	67.81	1.38203	1	2	2	12.	4.7378	18.9	20398.2
24	68.21	1.37488	2	0	3	6.	4.6831	30.3	25876.0
25	68.21	1.37488	0	2	3	6.	4.6831	15.8	7005.2
26	68.38	1.37189	3	0	1	6.	4.6604	1.4	54.5
27	68.38	1.37189	0	3	1	6.	4.6604	26.7	19933.8
28	73.54	1.28783	1	0	4	6.	4.0658	18.7	8570.9
29	73.54	1.28783	0	1	4	6.	4.0658	9.8	2331.3
30	75.74	1.25588	0	3	2	6.	3.8639	13.9	4479.0
31	75.74	1.25588	3	0	2	6.	3.8639	20.3	9565.8
32	77.75	1.22832	2	2	0	6.	3.7016	18.2	7334.1
33	79.97	1.19976	2	1	3	12.	3.5462	16.4	11512.9
34	79.97	1.19976	1	2	3	12.	3.5462	8.1	2778.1
35	80.13	1.19778	2	2	1	12.	3.5359	9.9	4148.3
36	81.26	1.18394	1	1	4	12.	3.4660	17.3	12441.7
37	81.58	1.18013	3	1	0	12.	3.4474	18.0	13456.2
38	83.93	1.15297	3	1	1	12.	3.3225	7.3	2150.1
39	83.93	1.15297	1	3	1	12.	3.3225	13.0	6734.5
40	85.05	1.14061	0	2	4	6.	3.2706	1.5	42.3
41	85.05	1.14061	2	0	4	6.	3.2706	8.1	1292.3
42	87.17	1.11823	2	2	2	12.	3.1852	1.8	122.0
43	87.54	1.11443	3	0	3	6.	3.1719	7.5	1058.8
44	87.54	1.11443	0	3	3	6.	3.1719	2.8	153.9
45	90.93	1.08151	3	1	2	12.	3.0717	10.4	3975.4
46	90.93	1.08151	1	3	2	12.	3.0717	16.2	9726.8
47	92.89	1.06376	4	0	0	6.	3.0305	11.9	2568.9
48	94.76	1.04769	0	1	5	6.	3.0019	17.8	5686.2
49	94.76	1.04769	1	0	5	6.	3.0019	10.3	1892.8
50	95.23	1.04373	0	4	1	6.	2.9963	7.1	468.1
51	95.23	1.04373	4	0	1	6.	2.9963	16.0	4593.2
52	96.35	1.03454	1	2	4	12.	2.9854	2.9	310.4
53	96.35	1.03454	2	1	4	12.	2.9854	14.4	7413.9
54	98.06	1.01489	2	2	3	12.	2.9736	13.4	6425.0

OBSERVED PEAKS AND MILLER PLANES UNDER EACH PEAK

PEAK NO	MILLER PLANES (PLANE NO) UNDER PEAK	I(CALC)	I(OBS)	SIGMA	CORRELATIONS TO NEXT PEAKS (%)
1	1	89192	2932	62	
2	2 3	452695	14110	122	
3	4	32332	1034	45	
4	5 6	32557	927	43	
5	7	14916	465	38	
6	8	23795	699	41	
7	9 10	14175	486	38	
8	11	58970	1830	70	-48
9	12	1448	35	60	
10	13 14	17676	557	39	-24
11	15 16	7980	212	34	
12	17	1079	31	32	
13	18 19	43131	1313	48	
14	20	8130	226	35	
15	21	2045	59	11	
16	22 23	26791	823	45	-19 2
17	24 25	32881	947	62	-36
18	26 27	19987	568	50	
19	28 29	10901	283	35	
20	30 31	14044	409	37	
21	32	7334	223	34	
22	33 34	14290	428	53	-43
23	35	4148	95	51	
24	36	12441	335	53	-41
25	37	13456	389	55	
26	38 39	8884	273	35	
27	40 41	1334	37	32	
28	42	122	3	31	
29	43 44	1211	36	32	
30	45 46	13701	451	38	
31	47	2568	83	32	
32	48 49	7578	205	34	
33	50 51	5061	151	33	
34	52 53	7723	234	35	
35	54	6425	210	34	

THE NUMBER OF OBSERVED PEAKS INCLUDED IS   35

FORM FACTORS AND MILLER INDICES, INCLUDING SPACE GROUP EXTINCTIONS ARE TAKEN FROM PROGRAM

THE NUMBER OF PLANES PROCESSED IN FUNNY IS   54

NUMBER CODES, VALUES AND STARTING INCREMENTS OF PARAMETERS

CODE NO	PARAMETER	VALUE	INCREMENT
1	SF SI	2.0000	0.0010
2	OC SI	3.0000	0.0010
3	X SI	0.4697	0.0010
4	Y SI	0.0	0.0010
5	Z SI	0.1667	0.0010
6	B11 SI	0.0053	0.0010
7	B22 SI	0.0078	0.0010
8	B33 SI	0.0049	0.0010
9	B12 SI	0.0039	0.0010
10	B13 SI	-0.0000	0.0010
11	B23 SI	-0.0000	0.0010
12	SF O	1.0000	0.0010
13	OC O	6.0000	0.0010
14	X O	0.4135	0.0010
15	Y O	0.2664	0.0010
16	Z O	0.2858	0.0010
17	B11 O	0.0190	0.0010
18	B22 O	0.0144	0.0010
19	B33 O	0.0083	0.0010
20	B12 O	0.0106	0.0010
21	B13 O	-0.0025	0.0010
22	B23 O	-0.0035	0.0010
23	SCALE	1.0000	0.0010
24	OTEMP	0.0	0.0010
25	PREFOR	0.0	0.0010

NO	I(CALC)	I (OBS)	NO	I(CALC)	I (OBS)	NO	I(CALC)	I (OBS)	NO	I(CALC)	I (OBS)	NO	I(CALC)	I (OBS)
1	89193.	2932.	11	7981.	212.	21	7334.	223.	31	2569.	83.	41	0.	0.
2	452697.	14110.	12	1080.	31.	22	14291.	428.	32	7579.	205.	42	0.	0.
3	32332.	1034.	13	43132.	1313.	23	4148.	95.	33	5061.	151.	43	0.	0.
4	32558.	927.	14	8131.	226.	24	12442.	335.	34	7724.	234.	44	0.	0.
5	14916.	465.	15	2045.	59.	25	13456.	389.	35	6425.	210.	45	0.	0.
6	23795.	699.	16	26792.	823.	26	8885.	273.	36	0.	0.	46	0.	0.
7	14176.	486.	17	32881.	947.	27	1335.	37.	37	0.	0.	47	0.	0.
8	58970.	1830.	18	19988.	568.	28	122.	3.	38	0.	0.	48	0.	0.
9	1448.	35.	19	10902.	283.	29	1213.	36.	39	0.	0.	49	0.	0.
10	17677.	557.	20	14045.	409.	30	13702.	451.	40	0.	0.	50	0.	0.

R - VALUE = 31.5098   (BASED ON UNWEIGHTED ABSOLUTE RESIDUALS)

```
STRUCTURE FACTOR AND LEAST SQUARES PROGRAM FOR OVERLAPPING POWDER DATA.

QUARZ COLLECTED WITH A GRAPHIT MONOCHROMATOR (PUBLISHED IN /5/)

LEAST SQUARES. CYCLE  1 OF  4 CYCLES.

REF$       PARAMETER       OLD VALUE      CHANGE         NEW VALUE      SIGMA

NO     1   SF  SI          2.00000        0.0            2.00000        0.0
NO     2   OC  SI          3.00000        0.0            3.00000        0.0
YES    3   X   SI          0.46970        0.00009        0.46979        0.00002
NO     4   Y   SI          0.0            0.0            0.0            0.0
NO     5   Z   SI          0.16667        0.0            0.16667        0.0
YES    6   B11 SI          0.00930        0.00032        0.00962        0.00006
YES    7   B22 SI          0.00780       -0.00009        0.00771        0.00008
YES    8   B33 SI          0.00490       -0.00004        0.00486        0.00004
NO     9   B12 SI          0.00390        0.0            0.00390        0.0
YES   10   B13 SI         -0.00001        0.00000       -0.00001        0.00002
NO    11   B23 SI         -0.00002        0.0           -0.00002        0.0
NO    12   SF  O           1.00000        0.0            1.00000        0.0
NO    13   OC  O           6.00000        0.0            6.00000        0.0
YES   14   X   O           0.41350        0.00013        0.41363        0.00003
YES   15   Y   O           0.26690       -0.00001        0.26689        0.00003
YES   16   Z   O           0.28577        0.00002        0.28579        0.00002
YES   17   B11 O           0.01900       -0.00022        0.01878        0.00015
YES   18   B22 O           0.01440       -0.00028        0.01412        0.00008
YES   19   B33 O           0.00830       -0.00002        0.00828        0.00014
YES   20   B12 O           0.01060       -0.00020        0.01040        0.00010
YES   21   B13 O          -0.00250        0.00005       -0.00245        0.00006
YES   22   B23 O          -0.00350        0.00023       -0.00327        0.00006
YES   23   SCALE          1.00000       -0.96854        0.03146        0.00053
NO    24   OTEMP          0.0            0.0            0.0            0.0
YES   25   PREFOR         0.0            0.00348        0.00348        0.00134

RWR/(N-M) OF LINEARIZED EQUATIONS =   .100856

RWR/(N-M) OF NONLINEAR  EQUATIONS =   .108386E+07


NO  I(CALC)  I (OBS)   NO  I(CALC)  I (OBS)   NO  I(CALC)  I (OBS)   NO  I(CALC)  I (OBS)   NO  I(CALC)  I (OBS)

 1   2810.    2932.    11   250.     212.     21   230.     223.     31    81.      83.     41    0.       0.
 2  14232.   14110.    12    34.      31.     22   448.     428.     32   237.     205.     42    0.       0.
 3   1017.    1034.    13  1354.    1313.     23   129.      95.     33   159.     151.     43    0.       0.
 4   1020.     927.    14   255.     226.     24   390.     335.     34   243.     234.     44    0.       0.
 5    469.     465.    15    64.      59.     25   423.     389.     35   202.     210.     45    0.       0.
 6    747.     699.    16   841.     823.     26   279.     273.     36     0.       0.     46    0.       0.
 7    447.     486.    17  1031.     947.     27    42.      37.     37     0.       0.     47    0.       0.
 8   1852.    1830.    18   627.     568.     28     4.       3.     38     0.       0.     48    0.       0.
 9     45.      35.    19   341.     283.     29    38.      36.     39     0.       0.     49    0.       0.
10    555.     557.    20   440.     409.     30   431.     451.     40     0.       0.     50    0.       0.

R - VALUE   =    0.0350      (BASED ON UNWEIGHTED ABSOLUTE RESIDUALS)
```

STRUCTURE FACTOR AND LEAST SQUARES PROGRAM FOR OVERLAPPING POWDER DATA.

QUARZ COLLECTED WITH A GRAPHIT MONOCHROMATOR (PUBLISHED IN /5/)

FUNCTIONS OF INTENSITIES

PEAK NO	MILLER PLANES UNDER PEAK	I(CALC)	I(OBS)	DELTA	DELTA/SIGMA	DELTA IN %
1	(1 0 0)	2926.62	2932.00	5.38	0.09	0.18
2	(1 0 1) , (0 1 1)	14124.36	14110.00	-14.36	-0.12	-0.10
3	(1 1 0)	1039.61	1034.00	-5.61	-0.12	-0.54
4	(1 0 2) , (0 1 2)	921.83	927.00	5.17	0.12	0.56
5	(1 1 1)	457.63	465.00	7.37	0.19	1.59
6	(2 0 0)	707.45	699.00	-8.45	-0.21	-1.21
7	(2 0 1) , (0 2 1)	490.49	486.00	-4.49	-0.12	-0.92
8	(1 1 2)	1794.99	1830.00	35.01	0.50	1.91
9	(0 0 3)	31.46	35.00	3.54	0.06	10.11
10	(2 0 2) , (0 2 2)	543.67	557.00	13.33	0.34	2.39
11	(1 0 3) , (0 1 3)	218.29	212.00	-6.29	-0.19	-2.97
12	(2 1 0)	35.66	31.00	-4.66	-0.15	-15.03
13	(2 1 1) , (1 2 1)	1313.47	1313.00	-0.47	-0.01	-0.04
14	(1 1 3)	240.92	226.00	-14.92	-0.43	-6.60
15	(3 0 0)	62.56	59.00	-3.56	-0.32	-6.04
16	(2 1 2) , (1 2 2)	807.41	823.00	15.59	0.35	1.89
17	(2 0 3) , (0 2 3)	948.87	947.00	-1.87	-0.03	-0.20
18	(3 0 1) , (0 3 1)	584.62	568.00	-16.63	-0.33	-2.93
19	(1 0 4) , (0 1 4)	288.37	283.00	-5.37	-0.15	-1.90
20	(0 3 2) , (3 0 2)	400.16	409.00	8.84	0.24	2.16
21	(2 2 0)	225.49	223.00	-2.49	-0.07	-1.12
22	(2 1 3) , (1 2 3)	417.11	428.00	10.89	0.21	2.54
23	(2 2 1)	103.75	95.00	-8.75	-0.17	-9.21
24	(1 1 4)	345.51	335.00	-10.51	-0.20	-3.14
25	(3 1 0)	405.85	389.00	-16.85	-0.31	-4.33
26	(3 1 1) , (1 3 1)	283.73	273.00	-10.73	-0.31	-3.93
27	(0 2 4) , (2 0 4)	44.06	37.00	-7.06	-0.22	-19.08
28	(2 2 2)	8.94	3.00	-5.94	-0.19	-198.05
29	(3 0 3) , (0 3 3)	48.03	36.00	-12.03	-0.38	-33.42
30	(3 1 2) , (1 3 2)	431.48	451.00	19.52	0.51	4.33
31	(4 0 0)	83.29	83.00	-0.29	-0.01	-0.36
32	(0 1 5) , (1 0 5)	211.85	205.00	-6.85	-0.20	-3.34
33	(0 4 1) , (4 0 1)	155.56	151.00	-4.56	-0.14	-3.02
34	(1 2 4) , (2 1 4)	232.56	234.00	1.44	0.04	0.61
35	(2 2 3)	212.49	210.00	-2.49	-0.07	-1.19

R - VALUE = 0.0100   (EASED ON WEIGHTED SQUARED RESIDUALS, WITHOUT CORRELATIONS))

R - VALUE = 0.0102   (CORRELATIONS INCLUDED)

```
STRUCTURE FACTOR AND LEAST SQUARES PROGRAM FOR OVERLAPPING POWDER DATA.

QUARZ COLLECTED WITH A GRAPHIT MONOCHROMATOR (PUBLISHED IN /5/)

LEAST SQUARES. CYCLE  4 OF  4 CYCLES.
```

REF$		PARAMETER	OLD VALUE	CHANGE	NEW VALUE	SIGMA
NO	1	SF SI	2.00000	0.0	2.00000	0.0
NO	2	OC SI	3.00000	0.0	3.00000	0.0
YES	3	X SI	0.47273	0.00001	0.47275	0.00077
NO	4	Y SI	0.0	0.0	0.0	0.0
NO	5	Z SI	0.16667	0.0	0.16667	0.0
YES	6	B11 SI	0.01899	0.00004	0.01904	0.00218
YES	7	B22 SI	0.00600	-0.00003	0.00597	0.00251
YES	8	B33 SI	0.00435	-0.00001	0.00435	0.00156
NO	9	B12 SI	0.00300	0.0	0.00300	0.0
YES	10	B13 SI	-0.00021	-0.00001	-0.00021	0.00076
NO	11	B23 SI	-0.00042	0.0	-0.00042	0.0
NO	12	SF O	1.00000	0.0	1.00000	0.0
NO	13	OC O	6.00000	0.0	6.00000	0.0
YES	14	X O	0.41614	0.00001	0.41615	0.00089
YES	15	Y O	0.26658	-0.00000	0.26658	0.00094
YES	16	Z O	0.28641	-0.00000	0.28641	0.00068
YES	17	B11 O	0.01594	-0.00002	0.01592	0.00502
YES	18	B22 O	0.00679	-0.00006	0.00672	0.00251
YES	19	B33 O	0.00579	0.00001	0.00580	0.00444
YES	20	B12 O	0.00721	-0.00002	0.00719	0.00337
YES	21	B13 O	0.00038	-0.00000	0.00038	0.00211
YES	22	B23 O	0.00443	0.00002	0.00444	0.00188
YES	23	SCALE	0.03221	-0.00000	0.03221	0.00056
NO	24	OTEMP	0.0	0.0	0.0	0.0
YES	25	PREFOR	0.11727	0.00002	0.11728	0.04317

```
RWR/(N-M) OF LINEARIZED EQUATIONS =  .115784

RWR/(N-M) OF NONLINEAR  EQUATIONS =  .115793
```

NO	I(CALC)	I (OBS)	NO	I(CALC)	I (OBS)	NO	I(CALC)	I (OBS)	NO	I(CALC)	I (OBS)	NO	I(CALC)	I (OBS)
1	2927.	2932.	11	218.	212.	21	225.	223.	31	83.	83.	41	0.	0.
2	14124.	14110.	12	36.	31.	22	417.	428.	32	212.	205.	42	0.	0.
3	1040.	1034.	13	1313.	1313.	23	104.	95.	33	156.	151.	43	0.	0.
4	922.	927.	14	241.	226.	24	346.	335.	34	233.	234.	44	0.	0.
5	458.	465.	15	63.	59.	25	406.	389.	35	212.	210.	45	0.	0.
6	707.	699.	16	807.	823.	26	284.	273.	36	0.	0.	46	0.	0.
7	490.	486.	17	949.	947.	27	44.	37.	37	0.	0.	47	0.	0.
8	1795.	1830.	18	585.	568.	28	9.	3.	38	0.	0.	48	0.	0.
9	31.	35.	19	288.	283.	29	48.	36.	39	0.	0.	49	0.	0.
10	544.	557.	20	400.	409.	30	431.	451.	40	0.	0.	50	0.	0.

```
R - VALUE  =   0.0097      (BASED ON UNWEIGHTED ABSOLUTE RESIDUALS)
```

APPENDIX A : NUMBER CODES OF LAUE GROUPS
              AND NONEXTICTION INDICATORS

The code word LAUE has the following interpretation:

    LAUE = 1   Laue group -1
    LAUE = 2   Laue group 2/m   (y axis unique)
    LAUE = 3   Laue group 2/m   (z axis unique)
    LAUE = 4   Laue group mmm
    LAUE = 5   Laue group 4/m
    LAUE = 6   Laue group 4/mmm
    LAUE = 7   Laue group m3
    LAUE = 8   Laue group m3m
    LAUE = 9   Laue group -3   (rhombohedral axes)
    LAUE =10   Laue group -3m   (rhombohedral axes)
    LAUE =11   Laue group -3   (hexagonal axes)
    LAUE =12   Laue group -3m1   (hexagonal axes)
    LAUE =13   Laue group 6/m
    LAUE =14   Laue group 6/mmm
    LAUE =15   Laue group -31m   (hexagonal axes)

The nonextinction code words I1 to I13 have the following
interpretation:

    I1 is code word for lattice type as follows
    I1 =0 P primitive no conditions
        1 A centered   k+l=2n
        2 B centered   h+l=2n
        3 C centered   h+k=2n
        4 F centered   h+k,k+l,l+h=2n
        5 I centered   h+k+l=2n
        6 R obverse   -h+k+l=3n
        7 R reverse   h-k+l=3n
        8 H hexagonal h-k=3n
        9 same as 0

    I2 is code word for (hk0)
    I2 =0 no conditions
        1 a glide   h=2n
        2 b glide   k=2n
        3 n glide   h+k=2n
        4 d glide   h+k=4n (h=2n,k=2n)

    I3 is code word for (h0l)
    I3 =0 no conditions
        1 a glide   h=2n
        2 c glide   l=2n
        3 n glide   h+l=2n
        4 d glide   h+l=4n (h=2n,l=2n)

    I4 is code word for (0kl)
    I4 =0 no conditions
        1 b glide   k=2n
        2 c glide   l=2n
        3 n glide   k+l=2n
        4 d glide   k+l=4n (k=2n,l=2n)

```
I5  is code word for (hhl)
I5 =0  no conditions
    1  c(n) glide  l=2n
    2  d glide     2h+l=4n

I6  is code word for (h-hl)
I6 =0  no conditions
    1  c(n) glide  l=2n
    2  d glide     2h+l=4n

I7  is code word for (hkh)
I7 =0  no conditions
    1  b(n) glide  k=2n
    2  d glide     2h+k=4n

I8  is code word for (hk-h)
I8 =0  no conditions
    1  b(n) glide  k=2n
    2  d glide     2h+k=4n

I9  is code word for (hkk)
I9 =0  no conditions
    1  a(n) glide  h=2n
    2  d glide     h+2k=4n

I10 is code word for (hk-k)
I10=0  no conditions
    1  a(n) glide  h=2n

I11 is code word for (h00)
I11=0  no conditions
    1  2(1) or 4(2) screw  h=2n
    2  4(1) or 4(3) screw  h=4n

I12     is code word for (0k0)
I12=0  no conditions
    1  2(1) or 4(2) screw  k=2n
    2  4(1) or 4(3) screw  k=4n

I13 is code word for (001)
I13=0  no conditions
    1  2(1) or 4(2) screw  l=2n
    2  3(1),3(2),6(2) or 6(4) screw  l=3n
    3  4(1) or 4(3) screw  l=4n
    4  6(1) or 6(5) screw  l=6n
```

APPENDIX B: NUMBER CODES OF FORM FACTOR TABLES ON TAPE

Atom	Origin	Code	Atom	Origin	Code
H	SCF	1	V+1	SCF	57
H-1	SCF	2	V+2	SCF	58
HE	SCF	3	V+3	SCF	59
HE-1	SCF	4	CR	SCF	60
LI	SCF	5	CR+1	SCF	61
LI+1	SCF	6	CR+2	SCF	62
LI-1	SCF	7	CR+3	SCF	63
BE	SCF	8	MN	SCF	64
BE+1	SCF	9	MN+1	SCF	65
BE+2	SCF	10	MN+2	SCF	66
B	SCF	11	MN+3	SCF	67
B+1	SCF	12	MN+4	SCF	68
B+2	SCF	13	FE	SCF	69
B+3	SCF	14	FE+1	SCF	70
C	SCF	15	FE+2	SCF	71
C+3	SCF	16	FE+3	SCF	72
C+4	SCF	17	FE+4	SCF	73
N	SCF	18	CO	SCF	74
N+3	SCF	19	CO+1	SCF	75
N+4	SCF	20	CO+2	SCF	76
O	SCF	21	CO+3	SCF	77
OX-2	SUZK	22	NI	SCF	78
OH-1	SUZK	23	NI+1	SCF	79
H2O	SUZK	24	NI+2	SCF	80
F-1	SCF	25	NI+3	SCF	81
NE	SCF	26	CU	SCF	82
NA	SCF	27	CU+1	SCF	83
NA+1	SCF	28	CU+2	SCF	84
MG	SCF	29	CU+3	SCF	85
MG+2	SCF	30	ZN	SCF	86
AL	SCF	31	ZN+2	SCF	87
AL+2	SCF	32	GA	SCF	88
AL+3	SCF	33	GA+1	SCF	89
SI	SCF	34	GA+3	SCF	90
CI+4	SCF	35	GE	SCF	91
P	SCF	36	GE+2	SCF	92
S	SCF	37	GE+4	SCF	93
S-1	SCF	38	CA	TFDS	94
S-2	SCF	39	CA+1	TFDS	95
CL	SCF	40	CA+2	TFDS	96
CL-1	SCF	41	SC	TFDS	97
A	SCF	42	SC+1	TFDS	98
K	SCF	43	SC+2	TFDS	99
K+1	SCF	44	SC+3	TFDS	100
CA	SCF	45	TI	TFDS	101
CA+1	SCF	46	TI+1	TFDS	102
CA+2	SCF	47	TI+2	TFDS	103
SC	SCF	48	TI+3	TFDS	104
SC+1	SCF	49	V	TFDS	105
SC+2	SCF	50	V+1	TFDS	106
SC+3	SCF	51	V+2	TFDS	107
TI	SCF	52	V+3	TFDS	108
TI+1	SCF	53	CR	TFDS	109
TI+2	SCF	54	CR+1	TFDS	110
TI+3	SCF	55	CR+2	TFDS	111
V	SCF	56	CR+3	TFDS	112

Atom	Origin	Code	Atom	Origin	Code
MN	TFDS	113	SB	TFDS	168
MN+1	TFDS	114	TE	TFDS	169
MN+2	TFDS	115	I	TFDS	170
MN+3	TFDS	116	XE	TFDS	171
MN+4	TFDS	117	CS	TFDS	172
FE	TFDS	118	BA	TFDS	173
FE+1	TFDS	119	LA	TFDS	174
FE+2	TFDS	120	CE	TFDS	175
FE+3	TFDS	121	PR	TFDS	176
FE+4	TFDS	122	ND	TFDS	177
CO	TFDS	123	PM	TFDS	178
CO+1	TFDS	124	SM	TFDS	179
CO+2	TFDS	125	EU	TFDS	180
CO+3	TFDS	126	GD	TFDS	181
NI	TFDS	127	TB	TFDS	182
NI+1	TFDS	128	DY	TFDS	183
NI+2	TFDS	129	HO	TFDS	184
NI+3	TFDS	130	ER	TFDS	185
CU	TFDS	131	TM	TFDS	186
CU+1	TFDS	132	YB	TFDS	187
CU+2	TFDS	133	LU	TFDS	188
CU+3	TFDS	134	HF	TFDS	189
ZN	TFDS	135	TA	TFDS	190
ZN+2	TFDS	136	W	TFDS	191
GA	TFDS	137	RE	TFDS	192
GA+1	TFDS	138	OS	TFDS	193
GA+3	TFDS	139	IR	TFDS	194
GE	TFDS	140	PT	TFDS	195
GE+2	TFDS	141	AU	TFDS	196
GE+4	TFDS	142	AU+1	TFDS	197
AS	TFDS	143	HG	TFDS	198
AS+1	TFDS	144	HG+2	TFDS	199
AS+2	TFDS	145	TL	TFDS	200
AS+3	TFDS	146	TL+1	TFDS	201
SE	TFDS	147	TL+3	TFDS	202
BR	TFDS	148	PB	TFDS	203
KR	TFDS	149	PB+3	TFDS	204
RB	TFDS	150	BI	TFDS	205
RB+1	TFDS	151	PO	TFDS	206
SR	TFDS	152	AT	TFDS	207
Y	TFDS	153	RN	TFDS	208
ZR	TFDS	154	FR	TFDS	209
ZR+4	TFDS	155	RA	TFDS	210
NB	TFDS	156	AC	TFDS	211
MO	TFDS	157	TH	TFDS	212
MO+1	TFDS	158	PA	TFDS	213
TC	TFDS	159	U	TFDS	214
RU	TFDS	160	NP	TFDS	215
RH	TFDS	161	PU	TFDS	216
PD	TFDS	162	AM	TFDS	217
AG	TFDS	163	CM	TFDS	218
AG+1	TFDS	164	BK	TFDS	219
CD	TFDS	165	CF	TFDS	220
IN	TFDS	166	ES	TFDS	221
SN	TFDS	167			

Index

GPSR Compliance
The European Union's (EU) General Product Safety Regulation (GPSR) is a set
of rules that requires consumer products to be safe and our obligations to
ensure this.

If you have any concerns about our products, you can contact us on

ProductSafety@springernature.com

In case Publisher is established outside the EU, the EU authorized
representative is:

Springer Nature Customer Service Center GmbH
Europaplatz 3
69115 Heidelberg, Germany